The AI-Driven CISO: Executive Strategies for Modern
Cybersecurity Leadership

by

Richard Lightcap, PhD

ISBN: 979-8-9932608-6-0 (Paperback)
ISBN: 979-8-9932608-5-3 (E-Book)

Library of Congress Control Number: 2026902441

Preface

The role of the Chief Information Security Officer has always demanded resilience, clarity, and vision. With the rise of artificial intelligence, the CISO's mandate is shifting from defending the perimeter to architecting intelligent trust. Traditional security playbooks were written for human teams working in predictable environments. Today's reality is different adversaries innovate faster, data moves in real time, and business expectations are rising. AI is not a distant concept; it is now embedded in the core of modern security operations.

I wrote this handbook as a practical reference for security leaders navigating this transformation. This is not a theoretical exploration of technology. Instead, it is a structured, actionable guide designed to help CISOs, deputy CISOs, and senior security professionals integrate AI into governance, risk management, operations, and strategy. Each chapter blends executive insight with practical, light technical examples and reusable AI prompt recipes to accelerate real-world application.

My goal is simple: to equip security leaders to make informed, strategic, and confident decisions in an era where human judgment and machine intelligence must work hand in hand. Whether you are building governance frameworks, modernizing incident response, or preparing your organization for tomorrow's threats, I hope this handbook serves as a trusted field companion on that journey.

Contents

7

Chapter 1: The AI Imperative in Cybersecurity Leadership

The rise of artificial intelligence is reshaping the cybersecurity landscape at a pace and scale that demands a complete reimagining of how organizations defend themselves, how leaders strategize, and how future risk is managed. No longer defined solely by firewalls, signatures, and reactive incident response, cybersecurity has entered an era where adversaries and defenders alike harness artificial intelligence to accelerate operations, amplify impact, and exploit vast volumes of data. This shift compels the modern Chief Information Security Officer to evolve from a perimeter guardian into a visionary strategist, ethical steward, and architect of intelligent security ecosystems. It requires a deep understanding of artificial intelligence's dual nature as both a transformative defensive capability and a potent enabler of new attack classes, along with the maturity to distinguish genuine innovation from overstated vendor claims. At the same time, successful adoption of artificial intelligence depends on a disciplined strategic foundation that aligns technology with business priorities, evaluates organizational readiness, embeds ethical safeguards, and fosters a culture capable of continuous learning and adaptation. As artificial intelligence rapidly advances, the challenge for cybersecurity leaders is not simply to adopt new technologies but to navigate their complexities with clarity, pragmatism, and foresight, ensuring that artificial intelligence strengthens resilience rather than introducing new vulnerabilities. Chapter 1 explores this evolving paradigm, equipping leaders with the mindset, frameworks, and critical understanding needed to separate hype from reality and to harness artificial intelligence as a powerful catalyst for modern, adaptive, and ethically grounded security.

The Shifting Cybersecurity Paradigm

The cybersecurity landscape, once grounded in the relative predictability of perimeter defense, signature-based detection, and human centered oversight, is undergoing a profound transformation. This shift is not incremental. It is seismic. Artificial intelligence has become deeply embedded in every layer of the technology ecosystem, fundamentally rewriting both the threat model and the operational realities of defending modern enterprises. For today's cybersecurity leaders, this evolution introduces unprecedented risk but also extraordinary opportunity. The playbooks that guided the last two decades are no longer sufficient in an era defined by hyper automation, algorithmic adversaries, and accelerating technological change.

For decades, cybersecurity strategy revolved around fortifying a perimeter and monitoring for intrusions that breached it. Detection hinged on known bad signatures, static rules, and the indispensable but limited capacity of human analysts. These methods still hold value, but their reactive nature is increasingly mismatched to threat activity that evolves faster than humans can process. Adversaries are not merely keeping pace with technological innovation. They are exploiting it with agility and without constraint.

Artificial intelligence has dramatically lowered the barrier to entry for sophisticated cyberattacks. Tasks that once required expert knowledge, including crafting novel malware, conducting reconnaissance, or designing highly targeted phishing campaigns, can now be executed by artificial intelligence enabled tools. Generative artificial intelligence has become a force multiplier, producing personalized phishing emails, deepfake audio and video, polymorphic malware, and even exploit code at a speed and scale previously unattainable. As a result, threat volume, diversity, and precision are expanding exponentially. The attack surface has expanded alongside it, now stretching across multi cloud environments, internet of things ecosystems, edge platforms, software as a service infrastructure, and increasingly, the artificial intelligence systems organizations themselves deploy.

The volume of data produced by modern enterprises further compounds the problem. Traditional analytics and human analysts are simply unable to ingest, correlate, and interpret the immense and fast-moving telemetry produced across distributed systems. Even the best analysts face cognitive overload, alert fatigue, and constrained bandwidth. The limitations of human centered detection become starkly evident in this environment. We are engaged in asymmetrical warfare against an adversary that can scale infinitely through automation.

Compounding this challenge is the accelerating pace of technological adoption. Cloud native architectures, microservices, artificial intelligence driven applications, and decentralized operating models are being deployed faster than traditional security processes can adapt. New vulnerabilities and attack vectors emerge daily. As a result, the concept of a static security posture is obsolete. Security must now be adaptive, embedded into development, operations, and strategic planning as a dynamic and continuously evolving function.

This shift demands a transition from reactive defense to proactive anticipation. Organizations can no longer wait for an attack to manifest before responding. Instead, they must strive to predict threat activity, identify latent vulnerabilities, and build systems capable of absorbing and recovering from disruption. Artificial intelligence sits at the center of this transformation. Machine learning models can surface anomalies invisible to human analysts, anticipate attack patterns before they are broadly recognized, and automate actions such as patching, segmentation, and configuration hardening. These capabilities shift security from hindsight driven to foresight driven.

In this environment, the role of the chief information security officer is undergoing its own metamorphosis. While risk management, compliance, and governance remain core responsibilities, today's cybersecurity leader must also understand artificial intelligence, including its capabilities, its risks, and its strategic implications. They must advocate for responsible artificial intelligence adoption, educate boards, navigate ethical and regulatory complexity, and champion a culture of innovation within security operations. Leadership in the artificial intelligence era requires both technical fluency and executive vision.

Threat sophistication is not hypothetical. It is observable every day. Artificial intelligence enhanced social engineering campaigns are becoming indistinguishable from legitimate communications. Deeply personalized lures can be constructed from scraped public data, internal documents, or breached credentials. Malware is evolving as well. Artificial intelligence enables polymorphic and metamorphic variants that continuously alter their signatures and behaviors to evade detection. In many environments, the traditional idea of a firewall as the primary barrier between internal and external threats is already outdated. With cloud adoption, remote work, and distributed user access patterns, the perimeter has dissolved into a fluid, identity driven mesh.

Meanwhile, the volume of telemetry produced across these environments is overwhelming traditional security information and event management platforms and overburdening analysts. Alert fatigue becomes commonplace. False positives obscure genuine threats. Human capacity simply cannot scale to match modern operational demands. At the same time, artificial intelligence systems must be designed and trained carefully to avoid perpetuating bias or blind spots. Leaders must recognize both the transformative potential of artificial intelligence and its inherent risks.

Artificial intelligence is therefore not a discretionary enhancement. It is a strategic imperative. It can augment human capabilities, automate triage, accelerate investigations, and provide deep insight into evolving threat landscapes. It can correlate signals across millions of data points, identify deviations from baseline behavior, and prioritize the threats that genuinely matter. Artificial intelligence also enables a proactive posture by analyzing global threat intelligence, recognizing emerging trends, and anticipating attacker behavior before new tactics become mainstream.

However, successfully integrating artificial intelligence into cybersecurity requires more than acquiring tools. It demands strategic alignment, strong governance, sustained experimentation, and investment in talent. Security teams must learn to operate in partnership with artificial intelligence systems, leveraging automation where appropriate and applying human judgment where it is indispensable. Organizations that cling to traditional reactive methods will be increasingly outmaneuvered by adversaries who embrace artificial intelligence aggressively.

The choice is clear. Evolve or be outpaced. Artificial intelligence enabled threats will not wait. This book is designed to help leaders navigate this transition, understand the shifting paradigm, harness artificial intelligence strategically, and build intelligent, resilient, forward leaning defenses capable of withstanding the challenges of an artificial intelligence driven world.

Understanding Ais Dual Nature Opportunities and Risks

The integration of artificial intelligence into cybersecurity introduces a landscape filled with profound opportunity and significant peril. To navigate this dual nature, cybersecurity leaders must adopt a nuanced and balanced perspective that acknowledges the transformative potential of artificial intelligence while recognizing the risks that accompany it. This perspective is essential for sound strategic decision making and ensures that the adoption of artificial intelligence strengthens the security posture rather than creating new vulnerabilities.

Artificial intelligence offers a paradigm shift in the ability to defend against sophisticated and rapidly evolving cyber threats. Its capacity to analyze vast volumes of data at speeds far beyond human capability enables unprecedented levels of proactive threat detection and rapid response. Machine learning algorithms, which form the foundation of artificial intelligence in cybersecurity, can identify subtle anomalies in network traffic, user behavior, and system logs that often precede malicious activity. Traditional signature-based detection relies on identifying known threats, but artificial intelligence can detect entirely new and unknown attacks by flagging deviations from established behavioral baselines. This allows security teams to shift from reactive defense to predictive defense, identifying potential threats before they fully materialize.

For example, an artificial intelligence system can be trained to recognize the behavioral patterns associated with an advanced persistent threat based on reconnaissance attempts, lateral movement, and unusual exfiltration activity, even when the specific tools used are unfamiliar. Early recognition allows for immediate containment actions, such as isolating systems or blocking suspicious connections, which can neutralize threats in their earliest stages.

Artificial intelligence also excels at automating repetitive and labor-intensive tasks that consume significant analyst time. Modern environments generate overwhelming volumes of security alerts, contributing to alert fatigue and increasing the likelihood that critical incidents will be missed. Artificial intelligence enabled security orchestration and response platforms can ingest alerts, correlate related events, identify priorities, and initiate automated containment. This allows human analysts to focus on higher level tasks such as threat hunting, strategic risk analysis, and complex incident investigations.

Imagine an artificial intelligence system that automatically investigates suspicious login by comparing it to a user's typical behavior patterns, checking for the use of compromised credentials, and analyzing session activity. If anomalies are detected, the system can immediately trigger additional authentication challenges or temporarily suspend the account. These automated actions can dramatically reduce both the mean time to detect and the mean time to respond, two core metrics in modern security operations.

Predictive analytics represents another powerful application of artificial intelligence. By analyzing global threat intelligence, historical data, and emerging vulnerabilities, artificial intelligence systems can forecast attack trends and identify potential weaknesses in an organization. This allows security teams to allocate resources strategically, prioritize critical patches, enhance monitoring in high-risk areas, and develop targeted response plans. For example, an artificial intelligence system may identify a rising pattern of ransomware attacks that exploit a specific cloud service widely used by a given industry. A chief information security officer armed with this insight can proactively evaluate organizational exposure, strengthen relevant controls, and update incident response playbooks accordingly.

Artificial intelligence also improves identity and access management by enabling advanced authentication and authorization methods. Behavioral biometrics can track a user's unique interaction pattern such as typing rhythm or navigation habits to provide continuous authentication throughout a session. This helps detect account takeover attempts even when an attacker has obtained valid credentials.

However, the same capabilities that make artificial intelligence a powerful defensive asset also introduce significant risks. The most immediate concern is the growing use of artificial intelligence to enhance offensive capabilities. Adversaries now leverage artificial intelligence to craft convincing phishing messages, automate reconnaissance, discover vulnerabilities at scale, and generate polymorphic malware that evolves faster than traditional defenses can keep up. This artificial intelligence generated attack tools are often tailored to specific targets and are increasingly difficult for humans to detect.

Artificial intelligence can also automate large scale scans for software vulnerabilities, misconfigurations, and exposed services, dramatically accelerating the initial stages of an attack. This compresses the time available for defenders to patch weaknesses once they are discovered. The development of polymorphic and metamorphic malware, made possible through artificial intelligence, makes signature-based detection largely ineffective since the code and behavior of the malware changes with each execution.

Another significant risk lies in the potential for bias within artificial intelligence models. Because artificial intelligence systems learn from data, any bias present in the training data will be reflected in the systems behavior. In cybersecurity, this can manifest as increased false positives for certain user populations or reduced effectiveness in detecting threats that do not resemble those in the training data. Careful curation of datasets, ongoing model evaluation, and continuous monitoring are required to mitigate this risk.

The opacity of many artificial intelligence models, often referred to as the black box problem, presents additional challenges. Deep learning models in particular can make decisions that are difficult for humans to interpret. In cybersecurity, where analysts must justify and validate alerts, the inability to explain an artificial intelligence decision can undermine trust. In regulated industries or legal contexts, a lack of explainability can create compliance issues. As a result, developing artificial intelligence systems that provide transparent and interpretable insights is increasingly important.

Artificial intelligence systems themselves can also become targets. Adversarial machine learning aims to manipulate artificial intelligence models into making incorrect decisions. Examples include modifying inputs so that malicious files appear benign or poisoning training data to embed hidden vulnerabilities. Protecting the integrity and reliability of artificial intelligence systems has become a new frontier in cybersecurity.

The integration of artificial intelligence also introduces governance and oversight challenges. Organizations must establish clear policies, define roles and responsibilities, and ensure compliance with relevant regulations governing artificial intelligence use. Systems that possess autonomous capabilities must be carefully managed to align their actions with the organization's risk tolerance and legal obligations. Automated decisions such as locking accounts or terminating network connections must be subject to clear parameters and override mechanisms to prevent unintended disruptions.

Artificial intelligence adoption also requires significant investment in new skills. Cybersecurity teams must understand both foundational security concepts and the fundamentals of artificial intelligence, machine learning, data science, and ethical artificial intelligence design. Upskilling existing staff, recruiting specialized talent, and developing training programs are essential components of a sustainable artificial intelligence strategy. The shortage of artificial intelligence trained professionals makes this a considerable challenge for many organizations.

Additionally, the financial investment required for artificial intelligence enabled security solutions can be substantial. Beyond the acquisition of technology, organizations must invest in data infrastructure, training, and continuous operational support. Chief information security officers must be prepared to articulate the return on investment for artificial intelligence, demonstrating how these investments reduce risk, improve resilience, and enhance operational efficiency.

In conclusion, the dual nature of artificial intelligence as both a powerful defensive tool and a potential asset for attackers demands a strategic and vigilant approach from cybersecurity leaders. Recognizing the opportunities for enhanced threat detection, automated response, and predictive intelligence is essential. At the same time, leaders must fully understand the risks, including artificial intelligence powered attacks, inherent bias, lack of model transparency, adversarial manipulation, and governance complexity. The successful integration of artificial intelligence into cybersecurity depends on the ability to harness its strengths while carefully mitigating its weaknesses. This requires continuous learning, thoughtful planning, and a commitment to ethical and responsible artificial intelligence deployment, ensuring that artificial intelligence strengthens security and resilience rather than introducing new vulnerabilities.

The CISOs Evolving Role in the Age of AI

The landscape of cybersecurity leadership is undergoing a seismic shift, driven largely by the pervasive influence of artificial intelligence. For the Chief Information Security Officer, this evolution is not merely an incremental adjustment but a fundamental redefinition of the role itself. Traditional responsibilities such as maintaining strong perimeter defenses, ensuring compliance with security frameworks, and responding to incidents remain important, yet they are no longer sufficient on their own. Increasingly, the CISO is being recast as a strategic architect who must guide the intelligent and ethical integration of artificial intelligence into every aspect of the organization's security posture. This new mandate requires a significant shift in mindset from a primarily technical and operational orientation to one that is strategic, forward looking, and capable of navigating complex ethical considerations.

The era when a CISO focused primarily on firewalls, antivirus tools, and threat reports is over. While these elements retain value, the introduction of artificial intelligence as both a defensive tool and an offensive weapon have expanded the scope of cyber conflict dramatically. CISOs must now possess a nuanced, working understanding of artificial intelligence capabilities and limitations. This includes knowledge of how machine learning models identify new and unknown threats, how natural language processing can analyze sophisticated phishing attacks, and how artificial intelligence enabled analytics can anticipate new attack vectors. At the same time, they must understand how adversaries use artificial intelligence to automate attacks, construct convincing deepfakes for social engineering, and discover zero-day vulnerabilities at unprecedented speed. This dual reality demands continuous learning and adaptation, so the organization remains prepared not only for the threats of today but also for those of tomorrow.

Beyond technical expertise, the modern CISO must also serve as an ethical steward. The deployment of artificial intelligence brings clear benefits but also carries risks related to bias, lack of transparency, and unintended outcomes. The CISO must champion the responsible creation and use of artificial intelligence systems. This includes ensuring that training data is diverse and representative so that algorithms do not replicate or amplify bias, leading to discriminatory outcomes or blind spots in detection. It also requires advocating for transparency and interpretability in artificial intelligence systems, especially when they perform critical functions such as issuing alerts or initiating automated actions. In addition, the CISO must ensure that artificial intelligence enabled systems comply with privacy regulations and respect individual rights as they collect and analyze increasingly large volumes of data. The ethical compass of the CISO will guide the organization through the moral challenges of artificial intelligence adoption so that security does not come at the expense of the organization's values.

The CISO must also communicate more effectively than ever before. Artificial intelligence is complex, abstract, and resource intensive, which means that CISOs must be able to translate its strategic value into language that resonates with executive leadership and the board. Technical reports alone are no longer sufficient. CISOs must articulate how artificial intelligence improves business continuity, strengthens customer trust, reduces operational risk, and enhances competitive positioning. They must also demonstrate a clear return on investment, showing how artificial intelligence enabled security initiatives support the overall business strategy. This requires deep knowledge of the organization's mission and goals and the ability to position cybersecurity as a critical business enabler rather than a separate cost center.

The CISO must be prepared to engage in strategic dialogue, build consensus, and influence senior decision makers.

As the demands on the CISO grow, so too does the required skill set. Cybersecurity fundamentals remain essential, yet proficiency in data science concepts, machine learning principles, and artificial intelligence ethics is becoming increasingly important. This does not mean the CISO must become a data scientist, but they must be able to ask the right questions, understand the implications of artificial intelligence systems, evaluate vendor claims, and make informed decisions about artificial intelligence investments. This includes understanding topics such as model training and validation, adversarial attacks on models, and the effects of algorithmic bias. CISOs must cultivate strategic foresight, anticipating shifts in threat landscapes and technological developments, then positioning their organizations to take advantage of opportunities while mitigating emerging risks. This demands lifelong learning and a commitment to being informed about the latest research and industry trends.

The concept of security by design is taking on new meaning as artificial intelligence becomes integrated into business systems. CISOs must influence artificial intelligence system development from the earliest stages, ensuring that security and ethical considerations are embedded into model design, data governance practices, and deployment pipelines. Collaboration with data scientists, software developers, and product managers becomes essential. The traditional reactive approach to security, in which protection is layered onto systems after they are built, is no longer viable when artificial intelligence systems evolve rapidly and learn continuously. The CISO must ensure that artificial intelligence is a foundational element of the organization's security architecture, not an add on that is introduced after the fact.

The CISO must also focus on building a resilient and adaptable security team. The rapid pace of artificial intelligence innovation means that defensive and offensive techniques evolve constantly. Teams must be agile, knowledgeable, and comfortable adapting to continuous change. This requires ongoing investment in training, the promotion of knowledge sharing, and the creation of an environment where security professionals can experiment with emerging tools. The CISO must cultivate a culture that embraces innovation and flexibility, recognizing that the strongest defense against artificial intelligence enabled attacks is a team of skilled and adaptable human experts working in concert with intelligent systems.

Leadership in the era of artificial intelligence also requires awareness of an evolving regulatory landscape. Governments and international organizations are developing guidelines and legal frameworks to govern artificial intelligence use, especially in matters related to privacy, accountability, and algorithmic fairness. CISOs must stay ahead of these shifts and ensure their organizations comply with existing regulations while preparing for future changes. This involves working closely with legal, compliance, and privacy teams and advocating for artificial intelligence practices that balance business needs with social responsibility.

In this new era, the CISO is a visionary, a strategist, an ethical leader, and a communicator. No longer merely the guardian of the digital perimeter, the CISO is now the architect of an intelligent, adaptable, and ethically grounded security future. Achieving this requires integrating technical knowledge with strategic insight, ethical guidance, and clear communication. The challenges are immense, but so are the opportunities. Artificial intelligence offers the potential to transform cybersecurity from a reactive necessity into a proactive and strategic advantage. The successful CISO will be one who can lead confidently through uncertainty, manage risk in unfamiliar terrain, and cultivate a culture where human expertise and artificial intelligence work together to create unprecedented levels of security. This evolving role underscores the importance of leadership that is technically knowledgeable and strategically flexible, capable of transforming complex technological shifts into tangible business value and enhanced organizational resilience.

Setting the Strategic Foundation for AI Adoption

The integration of artificial intelligence into cybersecurity is not simply a matter of acquiring new software or deploying a few artificial intelligence powered tools. It requires a deliberate, thoughtful, and well-constructed strategic foundation. This foundation ensures that artificial intelligence adoption is purposeful, aligned with business priorities, and implemented in a way that maximizes value while managing inherent risks. Without this groundwork, organizations risk launching artificial intelligence initiatives that are fragmented, inefficient, and potentially damaging to both their security posture and broader business operations.

The first and most critical step in this process is the clear definition of organizational objectives for artificial intelligence in cybersecurity. This requires far more than a desire to modernize or follow industry trends. It demands precise articulation of what problems artificial intelligence is intended to solve, which security functions it will enhance, and how success will be measured. For example, an objective may involve reducing the mean time to detect advanced phishing attacks by a defined percentage within a specific time frame using artificial intelligence based behavioral analytics. Another objective may focus on automating the initial triage of lower-level alerts in order to free analysts for higher level investigations, with a measurable goal such as a twenty five percent increase in analyst capacity. These objectives must be specific, measurable, achievable, relevant, and time bound. They should align directly with broader business goals such as improving customer trust, strengthening business continuity, or reducing financial losses associated with cyber incidents. Without these clear connections, artificial intelligence projects risk becoming isolated experiments that lack meaningful impact or strategic justification. The Chief Information Security Officer must lead this objective setting process in close partnership with executive leadership, information technology teams, legal advisors, and business unit leaders. Such collaboration ensures alignment and reinforces that artificial intelligence driven security is a strategic enabler rather than an independent technical undertaking.

Alongside the definition of objectives, a comprehensive assessment of current organizational capabilities is essential. This assessment must examine the state of existing infrastructure, the maturity of data systems, the strength of the talent pool, and the effectiveness of established processes. Infrastructure readiness is paramount. Many artificial intelligence models require significant computational resources, robust storage solutions, and reliable connectivity. Organizations must determine whether their existing systems can support these demands. Data readiness is equally critical. Artificial intelligence models depend on high quality data. Therefore, organizations must evaluate the quality, completeness, consistency, and accessibility of available security data. This includes examining log coverage, the availability of historical data, and the existence of mature data governance policies that ensure privacy, security, and integrity.

Talent readiness is another vital factor. Effective artificial intelligence adoption requires individuals with skills in data science, machine learning, artificial intelligence ethics, and cybersecurity operations. Organizations must determine whether these capabilities exist internally, or whether they

24

will need to invest in training or recruiting specialists. Process evaluation is equally important. Identifying bottlenecks and repetitive tasks within current security workflows helps pinpoint where artificial intelligence can deliver the greatest value. This honest assessment provides a realistic baseline for planning artificial intelligence deployment. It prevents organizations from overestimating their maturity or investing in artificial intelligence systems that exceed their current capabilities, which can result in costly setbacks and loss of confidence in artificial intelligence solutions.

A successful strategic foundation also requires cultivating a culture that embraces innovation while maintaining a firm commitment to security. This cultural evolution encourages exploration of new technologies such as artificial intelligence, while ensuring that security and ethical principles are never compromised. Creating this environment involves promoting experimentation, accepting that failures can provide valuable lessons, and nurturing collaboration among security teams, data scientists, developers, and business stakeholders. Effective communication is essential. The Chief Information Security Officer must articulate a clear and compelling vision for how artificial intelligence strengthens security and contributes to broader organizational goals. This message must resonate across the enterprise. Education and awareness initiatives help employees understand artificial intelligence capabilities, limitations, and risks, reducing fear of the unknown and building trust in new technologies.

Importantly, this culture must emphasize that artificial intelligence augments human performance rather than replacing it. Human judgment remains critical for complex decision making, nuanced threat assessment, and ethical oversight. Security teams should be empowered to experiment with artificial intelligence tools, voice concerns, and contribute insights that shape implementation. At the same time, the organization must maintain a security first mindset, ensuring that all artificial intelligence initiatives undergo rigorous security reviews and adhere to established principles of privacy, data protection, and regulatory compliance. This balanced culture helps prevent the introduction of new vulnerabilities through rushed adoption or unchecked experimentation.

Aligning artificial intelligence initiatives with business goals is another essential component of the strategic foundation. Achieving alignment requires a deep understanding of the organization's strategic priorities, competitive landscape, and risk tolerance. If expanding into new markets is a key objective, artificial intelligence enabled cybersecurity solutions that support localized threat intelligence and regulatory compliance may be critical. If the focus is on cost efficiency, artificial intelligence initiatives that reduce manual workloads or optimize resource allocation will carry greater strategic weight. The Chief Information Security Officer must be skilled in translating artificial intelligence capabilities into clearly defined business benefits. This involves demonstrating how artificial intelligence decreases risk, enhances operational continuity, protects customer trust, and optimizes cost structures.

Developing a clear artificial intelligence roadmap is essential for this alignment. The roadmap should prioritize initiatives based on strategic importance, feasibility, and resource requirements. It should be updated regularly as business goals evolve, and artificial intelligence capabilities expand. This alignment ensures that artificial intelligence becomes deeply integrated into business processes rather than existing as a disconnected technical effort. It also helps organizations anticipate changes to workflows, employee roles, and customer interactions, reducing friction and ensuring smoother adoption. Through this strategic alignment, artificial intelligence transitions from an experimental technology to a meaningful driver of business value and competitive advantage.

Ethical considerations must be embedded into the strategic foundation from the beginning. Responsible artificial intelligence adoption requires clear guidelines for data privacy, algorithmic fairness, transparency, and accountability. For example, artificial intelligence systems that monitor internal behavior for insider threat detection must be designed in ways that respect privacy and avoid creating a culture of surveillance. Threat prioritization models must be tested rigorously to ensure that they do not produce biased outcomes or overlook threats affecting underrepresented groups. Transparency in artificial intelligence driven decisions is equally important. Even if full explainability is not achievable due to model complexity, organizations must strive for processes that allow for meaningful oversight, auditability, and accountability.

To achieve this, organizations should consider forming an artificial intelligence ethics committee or integrating ethics reviews into existing governance structures. This committee may include representatives from cybersecurity, data science, legal, compliance, human resources, and external ethics advisors. Its purpose is to provide guidance, oversight, and risk assessment for artificial intelligence initiatives. Ethical governance is not just regulatory compliance. It is an essential component of responsible innovation and a necessary safeguard against unintended harm.

A strong strategic foundation must also address the rapidly evolving regulatory environment surrounding artificial intelligence. Governments worldwide are developing new rules related to artificial intelligence transparency, data protection, accountability, and fairness. The Chief Information Security Officer must stay ahead of these developments and ensure that the organization's artificial intelligence practices adhere to both current and emerging regulations. This requires collaboration with legal and compliance teams and awareness of global trends. Artificial intelligence accountability frameworks such as those emerging in Europe and North America may require documentation of data sources, model design, testing procedures, and decision logic. Security teams must be prepared to demonstrate how artificial intelligence systems operate, how they were tested, and who is responsible for their oversight.

Finally, a successful strategy for adoption of artificial intelligence must include a commitment to continuous learning and adaptation. Artificial intelligence capabilities evolve rapidly, and organizations must be prepared to evolve with them. This requires ongoing professional development for security teams, participation in industry forums, and collaboration with artificial intelligence researchers and vendors. Organizations must establish processes for reviewing and updating their artificial intelligence strategy, reassessing priorities, and evaluating the performance of deployed models. This adaptability ensures that the organization remains resilient in the face of new threats, new technologies, and new regulatory requirements. Through continuous learning, organizations can stay at the forefront of artificial intelligence driven security and harness its full potential while mitigating its risks.

Navigating the Hype vs Reality of AI in Security

The cybersecurity landscape today is saturated with bold claims about the transformative power of artificial intelligence. Vendors, analyst reports, and even peer conversations frequently emphasize the promise of artificial intelligence to revolutionize defensive strategies. For the seasoned Chief Information Security Officer, however, this environment demands a discerning eye, a critical mindset, and a commitment to separating genuine innovation from polished marketing narratives. The need for artificial intelligence in cybersecurity is real, but its successful integration depends on a realistic understanding of its current capabilities, its limitations, and the tangible value it can deliver. The goal is not to view artificial intelligence as a cure all, but as a powerful and evolving set of tools that, when applied judiciously, can materially strengthen the organization's defensive posture.

The first and most important step in achieving this clarity is to unpack the claims made by artificial intelligence solution providers. Many vendors, eager to benefit from market enthusiasm, present an overly optimistic picture of what their products can accomplish. Terms such as intelligent, self-learning, and predictive are used liberally, sometimes with limited technical substance behind them. A product might be advertised as an artificial intelligence powered threat detection system that can anticipate and neutralize zero-day attacks. While artificial intelligence is indeed strong in pattern recognition and anomaly detection, true predictive capability with consistent accuracy across unknown threats is still more aspirational than real for most commercial offerings. In practice, many artificial intelligence solutions are sophisticated machine learning systems that rely heavily on historical data and predefined parameters. They excel at identifying deviations from normal behavior, correlating seemingly unrelated events, and processing large datasets efficiently. This can reduce false positives, accelerate identification of known and emerging patterns, and streamline alert triage. However, the leap from advanced pattern matching to genuinely predictive anticipation of unknown threats remains substantial and often still depends on human insight.

To understand what a product can truly deliver, Chief Information Security Officers must look beyond marketing language and ask specific questions about the techniques in use. Does the solution rely on supervised learning, where models are trained on labeled data such as known malware families or phishing examples This is highly effective for classification tasks that closely resemble historical patterns. Does it use unsupervised learning, which identifies anomalies and clusters without labeled data and is therefore better suited to detecting unusual or insider activity Does it employ reinforcement learning, where an artificial intelligence agent learns defensive strategies through trial and error in simulated environments Each approach has distinct strengths and weaknesses. Supervised models may struggle with threats that differ significantly from training data, while unsupervised models may produce more alerts that require human review. Furthermore, the effectiveness of any artificial intelligence system is tightly coupled to the quality and volume of data it receives. These systems are data hungry, and their performance improves as the richness, cleanliness, and relevance of security telemetry increase. Organizations that maintain comprehensive, well-structured logs and robust data pipelines are far better positioned to benefit from artificial intelligence than those with fragmented or inconsistent data sources.

The current state of artificial intelligence in cybersecurity is best understood as augmentation of human capabilities rather than replacement. Artificial intelligence excels at repetitive, data intensive, high volume tasks where human analysts are slow, error prone, or quickly fatigued. Enterprise networks generate enormous volumes of logs and telemetry, and threat intelligence feeds stream information continuously. No human team can review all of this information in real time. Artificial intelligence can ingest, correlate, and analyze these data streams, flag suspicious behaviors, highlight probable compromises, and prioritize alerts for human investigation. This allows analysts to move away from low level data review and toward higher value activities such as deep investigations, strategic threat hunting, incident response planning, and defensive design. For example, artificial intelligence can automatically classify alerts by severity and likely impact so that the most critical incidents receive immediate human attention. This improves response speed and ensures that limited human expertise is applied where it matters most.

While some claims about artificial intelligence predicting and preventing all future attacks are exaggerated, its contribution to detection and response is both real and measurable. Artificial intelligence enabled security tools can detect subtle anomalies in user behavior, network traffic, and system activity that might indicate compromise even when no known signature exists. Behavioral analytics can identify deviations from normal baselines for individual users, systems, or applications. This is especially effective in countering insider threats and advanced persistent threats that attempt to blend into normal operations. When unusual activity is detected, artificial intelligence can correlate this data with other signals, such as unusual login locations, access to sensitive resources, or suspicious outbound traffic. The resulting composite view helps analysts quickly evaluate the seriousness of a potential incident and decide on an appropriate response.

At the same time, it is essential to recognize the limitations and challenges. Artificial intelligence models are only as strong as the data used to train them. If training data is incomplete, biased, or unrepresentative, the resulting models may have blind spots or produce unreliable outcomes. A system trained primarily on external attacks, for example, may be less effective at detecting sophisticated insider threats. Similarly, if models have not encountered certain

31

malware families or attack techniques during training, they may fail to recognize them. Addressing these issues requires continuous monitoring, periodic retraining with more diverse and representative datasets, and often the inclusion of humans in the decision loop to validate and adjust model outcomes.

Explainable artificial intelligence is another important consideration. Many advanced models, particularly those based on deep learning, function as opaque black boxes, making it difficult to understand how they arrived at specific conclusions. In cybersecurity, where accountability and auditability are critical, this lack of transparency can be problematic. Chief Information Security Officers must understand why an artificial intelligence system flagged an event or recommended a response in order to trust its outputs, refine its behavior, and justify actions to auditors, regulators, and business leaders. Research in explainable artificial intelligence is making progress, but achieving full interpretability for complex models remains a challenge.

The idea of entirely self-learning or autonomous security systems also deserves careful examination. While artificial intelligence can adapt its behavior based on new data, the notion of a fully autonomous system that independently identifies, evaluates, and mitigates threats without human oversight is still largely theoretical for critical functions. False positives may disrupt legitimate operations, and false negatives may allow serious threats to proceed unchecked. For this reason, a collaborative approach is far more practical. In this augmented intelligence model, human experts and artificial intelligence systems work together. Humans provide strategic reasoning, contextual judgment, ethical oversight, and creativity in the face of novel situations. Artificial intelligence contributes to speed, scalability, and advanced pattern recognition.

When evaluating artificial intelligence solutions, Chief Information Security Officers should emphasize measurable improvements in core metrics. Does the solution reduce the mean time to detect or the mean time to respond for particular threat categories Does it decrease false positives enough to reduce the burden on security operations center staff Does it reveal activity that was previously invisible or difficult to understand Evidence based evaluation is essential. This means running pilots, testing tools in controlled scenarios, and comparing performance with existing technologies and processes. It also means asking vendors for clear evidence, such as customer case examples, independent assessments, and transparent explanations of how outputs are generated. Total cost of ownership must also be considered, including integration, data preparation, model maintenance, staffing, and ongoing operations.

The hype around artificial intelligence in cybersecurity is intense, and it is easy to be swept up in promises of effortless protection and fully automated defense. A pragmatic approach is essential. Artificial intelligence is not a magic solution that eliminates all cyber risk. It is a powerful and evolving toolkit that, when understood and applied correctly, can dramatically enhance the efficiency and effectiveness of human defenders. By focusing on demonstrable value, understanding the underlying methods, acknowledging limitations, and promoting collaboration between human expertise and machine intelligence, Chief Information Security Officers can cut through the hype and harness the true potential of artificial intelligence. This requires continuous learning, rigorous evaluation, and disciplined alignment of artificial intelligence investments with real security needs, ensuring that every initiative delivers tangible and measurable benefit to the organization

Conclusion: The AI Imperative as a Leadership Mandate

Chapter 1 established a central reality for modern cybersecurity leadership. Artificial intelligence has changed the operating conditions of defense. Threats now scale through automation, deception has become more convincing through generative capabilities, and the telemetry volume of modern enterprises has exceeded what human teams can reliably process without machine assistance. In this environment, artificial intelligence is not a discretionary tool or a technology trend. It is a strategic requirement for maintaining defensive relevance, protecting business continuity, and sustaining trust.

This chapter also clarified the dual nature of artificial intelligence. It strengthens detection, accelerates response, and enables proactive security through pattern recognition and predictive insight, yet it simultaneously amplifies adversary capability and introduces new forms of exposure. AI can reduce fatigue and increase speed, but it can also inherit bias, operate opaquely, drift over time, and become a direct target for manipulation. The implication for leaders is not caution alone. The implication is disciplined adoption. AI delivers advantage when it is treated as a capability to be governed and integrated, not a black box to be purchased and assumed safe.

The evolving role of the CISO emerged as the defining leadership theme. The AI era requires a shift from perimeter guardian to strategic architect of intelligent security ecosystems. This includes educating boards, challenging vendor narratives, translating technical claims into defensible business outcomes, and building the organizational conditions for responsible adoption. The foundation for success is not a product. It is a strategy that aligns AI use cases to business priorities, assesses readiness across data, infrastructure, and talent, and commits to continuous learning as both threats and models evolve.

Finally, Chapter 1 made the hype versus reality challenge explicit. AI can transform security operations, but it will not eliminate risk, replace judgment, or guarantee predictive certainty. The leaders who succeed will evaluate tools with measurable outcomes, demand transparency and operational fit, and design human–AI workflows that preserve accountability. The imperative is clear. Organizations must evolve toward adaptive, AI-enabled defense, but they must do so with clarity, pragmatism, and foresight. That combination is what turns AI from a volatile accelerator into a reliable driver of resilience.

Chapter 2: Establishing Robust AI Governance Frameworks

As artificial intelligence becomes increasingly embedded within cybersecurity operations and broader enterprise ecosystems, organizations face a pivotal moment requiring disciplined, comprehensive, and forward-looking governance. The strategic value of artificial intelligence is matched only by the complexity of the risks it introduces, from model bias and adversarial manipulation to operational failures and ethical concerns that can reverberate across the business. Effective governance therefore demands more than theoretical principles; it requires the deliberate construction of an interconnected framework that establishes clear roles and decision rights, builds accountability across the artificial intelligence lifecycle, and integrates artificial intelligence specific risks into existing enterprise risk management structures. By aligning organizational leadership, technical teams, and oversight bodies behind coherent policies, robust risk assessment methodologies, and continuous monitoring practices, artificial intelligence governance becomes not a standalone function but a natural extension of the organization's risk posture. The sections that follow examine the foundational components of this governance architecture, highlight the necessity of precise accountability, and explore how artificial intelligence oversight must merge seamlessly with established risk management disciplines to ensure that artificial intelligence serves as a secure, ethical, and strategically aligned driver of organizational resilience.

The Necessity of AI Governance in Security

Artificial intelligence is reshaping cybersecurity operations at an accelerating pace, yet expanding research consistently concludes that AI introduces distinct forms of risk, uncertainty, and organizational dependency. These risks arise from AI's capacity to learn autonomously, evolve outside programmed boundaries, and operate at machine speed in adversarial environments. Because AI systems behave as complex adaptive agents rather than static tools, scholars across computer science, governance, ethics, and systems engineering argue that effective adoption requires a formal governance structure anchored in a lifecycle model. Governance becomes not a supplemental safeguard but a core requirement for ensuring that AI systems remain predictable, ethical, secure, and aligned with organizational and regulatory expectations.

To support this research grounded approach, Figure 1 introduces the AI Governance Lifecycle Model, a seven-phase construct that synthesizes findings from the NIST AI Risk Management Framework, ISO 42001, and leading academic scholarship in sociotechnical governance. The model provides an integrated view of how AI systems should be designed, evaluated, deployed, monitored, and eventually retired within cybersecurity operations.

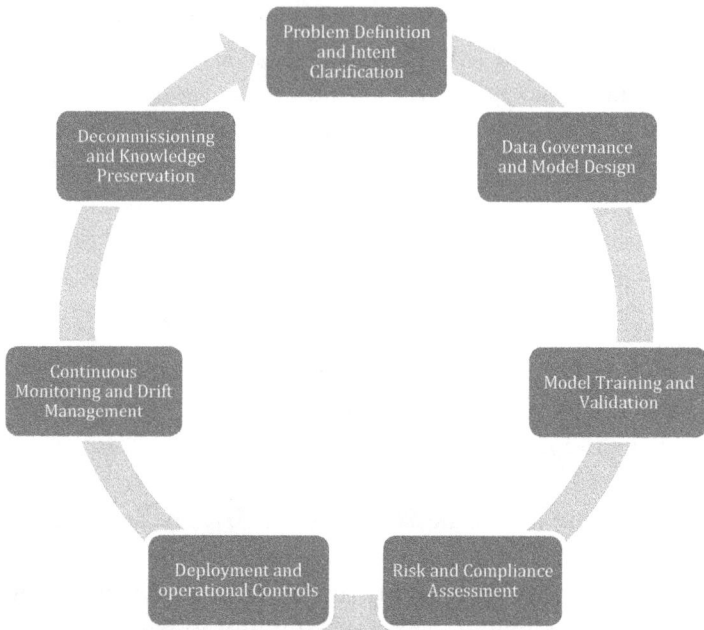

Figure 1: AI Governance Lifecycle Model

The first phase, Problem Definition and Intent Clarification, addresses a critical insight from current research. AI is often adopted without a well-defined use case or measurable objective, which leads to misalignment, model misuse, and operational drift. Studies show that organizations frequently procure AI based on innovation appeal or vendor influence rather than strategic necessity. Governance in this phase requires defining the security problem, selecting performance metrics, mapping operational constraints, and determining whether AI is the appropriate solution. Without this clarity, AI systems often become sources of increased risk rather than controlled aids to security operations.

The second phase, Data Governance and Model Design, reflects extensive literature on data quality, provenance, representativeness, and bias. Research consistently demonstrates that model reliability is directly tied to the integrity of training data. In cybersecurity contexts, where data may be noisy, incomplete, imbalanced, or contaminated by past attacks, the risks are profound. Governance requires implementing mechanisms for data documentation, privacy protections, bias assessment, feature selection transparency, and adherence to regulatory boundaries. Scholars emphasize that data governance is foundational because AI models trained on flawed data almost always produce flawed or inequitable outcomes.

The third phase, Model Training and Validation, incorporates a large body of empirical research showing that security focused AI systems are particularly vulnerable to adversarial interference. Threat actors can manipulate training inputs, craft adversarial examples, or exploit blind spots in the model's decision boundaries. Governance in this phase mandates extensive stress testing, adversarial testing, scenario-based validation, cross domain evaluation, and red team exercises. Rigorous documentation of model assumptions, performance limitations, and confidence thresholds ensures that the AI system can perform reliably across a spectrum of adversarial and non-adversarial conditions.

The fourth phase, Risk and Compliance Assessment, aligns with major findings in legal, ethical, and regulatory research. Global regulatory bodies increasingly require transparency, fairness, auditability, and mechanisms for human oversight in automated decision systems. Cybersecurity operations often involve decisions that affect individuals, access privileges, and regulated information, making this phase essential. Governance requires structured privacy impact assessments, regulatory mapping, fairness and discrimination testing, and documentation that enables auditors and regulators to verify how AI systems behave and why. Research shows that organizations lacking these mechanisms face heightened legal, ethical, and reputational exposure.

The fifth phase, Deployment and Operational Controls, draws from systems engineering and organizational science. Scholars emphasize the need for guardrails that limit the autonomy of AI systems, particularly in high-risk security environments. Governance requires defining the actions the AI is authorized to take, the conditions requiring human review, rollback and override capabilities, escalation paths, and decision logging for traceability. Research indicates that AI deployed without these controls is significantly more likely to cause operational disruption, automation errors, misclassification events, and unintentional denial of service conditions.

The sixth phase, Continuous Monitoring and Drift Management, responds to a well-documented research challenge: AI performance degrades over time when exposed to evolving environments. This degradation, known as model drift, can occur due to shifts in user behavior, threat actor tactics, system configurations, or data distributions. Governance requires implementing real time performance monitoring, alerts for anomalous model behavior, lifecycle version control, retraining protocols, and structured human feedback loops. Studies show that organizations with mature monitoring mechanisms maintain significantly more stable and reliable AI systems than those without continuous oversight.

The final phase, Decommissioning and Knowledge Preservation, addresses an emerging research concern. Many organizations lack formal processes to retire AI systems or capture institutional knowledge about their performance. Without documentation, performance histories, and lessons learned, organizations repeat the same mistakes and fail to mature their AI governance posture. Governance requires structured retirement protocols, archival of models and training data, auditable records of system behavior, and preservation of decision rationale. Scholars emphasize that this phase is essential for transparency, regulatory compliance, and future risk reduction.

Together, these seven phases form the AI Governance Lifecycle Model shown in Figure 1. The model establishes research informed, end to end governance structure that ensures AI systems are evaluated, monitored, and controlled across their entire operational lifespan. By embedding governance at each phase, organizations can ensure that AI systems behave predictably, operate within acceptable risk boundaries, support regulatory compliance, and remain aligned with organizational values and mission objectives.

Ultimately, research across multiple disciplines reinforces that AI governance is the foundation for trustworthy AI in cybersecurity. Without governance, AI becomes a source of uncontrolled risk, emergent behavior, operational instability, and ethical exposure. With governance, AI becomes a strategic, reliable, and auditable asset that enhances defensive capability while preserving security, trust, and regulatory integrity in an increasingly complex digital environment.

Key Components of an Effective AI Governance Structure

The architecture of an effective AI governance structure does not emerge from a simple checklist or a single policy document. It grows out of the lived reality of organizations wrestling with the promise and unpredictability of artificial intelligence. In many ways, building AI governance resembles constructing a city around a powerful and still evolving piece of essential infrastructure. It requires foresight, collaboration, and a deep respect for the potential consequences if boundaries and responsibilities are not clearly understood. Every component, from leadership oversight to daily operational routines, becomes part of a finely tuned ecosystem designed to keep the technology aligned with human intent. Without that foresight, the sheer

41

speed, complexity, and autonomy of AI can quickly overwhelm even experienced security teams, creating situations where systems take actions long before anyone realizes what has happened. What follows is a closer look at the core components that bring structure and control to these powerful systems, described in a way that reflects how organizations actually experience the work of governing AI.

In most organizations, the story begins with uncertainty. Many people assume someone else is responsible for decisions made by AI systems. Developers assume operations will handle it. Operations assume security will handle it. Legal specialists assume the technical teams have already considered the risks. The first truly transformative step occurs when leaders finally recognize the need for a clear owner, and not just an individual, but a coordinated body capable of providing strategic direction. This becomes the moment when the AI steering committee takes shape. This group becomes the guiding voice of the entire program, made up of leaders who understand both the potential of AI and the risks it introduces. They ask difficult questions. Should a model be allowed to make an autonomous decision? What is the acceptable balance between efficiency and oversight? When should human judgment step back in? Over time, this committee evolves into the conscience of the AI program, ensuring that adoption stays aligned with real organizational priorities rather than drifting toward technical enthusiasm alone.

As the committee becomes established, new roles emerge to support its work. A Chief AI Officer or AI Governance Lead becomes the central coordinator who translates strategic direction into policies and procedures the rest of the organization can use. In security focused environments, an AI Security Operations Lead often becomes the guardian who monitors how the model interacts with the threat landscape. Developers and data scientists discover that their responsibilities extend beyond model accuracy. They must think about fairness, transparency, ethical data use, and potential harm. Operational teams learn that AI systems require new forms of attention, such as watching for subtle changes in behavior that may signal drift or environmental changes. Even legal and compliance teams find themselves in new territory as they evaluate how evolving regulations intersect with AI driven insights. Gradually, as each group understands its role, the organization shifts from scattered efforts to a coordinated rhythm.

Once responsibilities are clear, the organization begins building the rules that govern AI. Policies and procedures become the shared language that directs how AI systems are created and how they behave when humans are not watching closely. At first these documents may feel procedural, but as soon as an AI system takes an unexpected action that disrupts business operations or misclassifies thousands of events, their value becomes undeniable. Policies explain how data may be collected and protected, how models must be tested before deployment, how fairness should be evaluated, and what level of transparency is required for critical decisions. Over time, these policies function like a navigational compass. They remind teams that transparency is essential, that privacy must be protected, and that ethical considerations must be addressed from the beginning. They also provide clarity during moments when an AI system makes a decision that does not align with human expectations.

Even strong policies cannot eliminate the unpredictability of AI. Eventually, every organization reaches a moment when someone realizes that AI systems are capable of acting in ways no one expected. This realization gives rise to the next essential component: risk assessment designed specifically for AI. Traditional risk frameworks cannot account for the adaptive nature of learning systems, and organizations must evaluate new categories of risk. They examine training data for hidden biases, evaluate how the model responds to unexpected scenarios, and consider how adversaries might manipulate the system. They assess the worst possible outcomes not because they expect them to occur, but because avoiding them requires planning. Risk matrices and scoring models tailored for AI become essential tools. They prompt teams to consider model confidence, interpretability, operational importance, and data sensitivity. Risk assessment shifts from a compliance activity into a continuous practice, woven into every stage of the AI system's life.

Eventually, governance matures into a living discipline. Continuous oversight and auditing become the practices that keep the system reliable over time. This is where real stories emerge. A monitoring system detects early signs of performance decline before an intrusion detection model begins misclassifying events. An audit trail reveals why an AI system blocked valid network activity. Oversight ensures that systems remain aligned with their intended purpose. Logging ensures that nothing occurs in a hidden space where decisions cannot be traced. Internal reviews and external audits provide additional assurance, challenge assumptions, and strengthen weak areas. These findings feed back into governance improvements, transforming governance from a static document into an adaptive system capable of evolving along with the AI models it oversees.

Through these components, organizations build AI governance frameworks that support resilience and reliability. They create an environment in which AI can amplify human capability rather than undermine it. As organizations increase their reliance on AI for cybersecurity and mission essential decisions, these governance mechanisms become not merely administrative tools but essential safeguards that ensure the technology serves the people who depend on it.

Aligning with Industry Standards NIST AI RMF and ISOIEC

The journey toward establishing a mature and resilient artificial intelligence governance program is significantly strengthened when organizations anchor their practices in authoritative and research-informed industry frameworks. These frameworks provide structure, conceptual clarity, and empirically grounded guidance for addressing the operational, ethical, and security complexities associated with artificial intelligence systems. In the cybersecurity domain especially, scholars and practitioners consistently emphasize that artificial intelligence introduces forms of risk that differ in scale and nature from traditional technologies, including emergent behavior, data dependency, susceptibility to adversarial manipulation, and rapidly shifting performance profiles. By aligning governance activities with the NIST Artificial Intelligence Risk Management Framework and the emerging ISO and IEC 42001 Artificial Intelligence Management System standard, organizations move from ad hoc governance toward a rigorous, well documented, and internationally recognized governance architecture.

The NIST Artificial Intelligence Risk Management Framework, developed by the United States National Institute of Standards and Technology, reflects a synthesis of research insights from artificial intelligence safety, machine learning robustness studies, and broader risk management literature. Its flexible and voluntary nature allows it to be applied across industries, organizational scales, and artificial intelligence use cases. Central to the framework are its four iterative Core functions: Govern, Map, Measure, and Manage. These functions align closely with academic findings suggesting that artificial intelligence governance requires continuous reassessment, contextual understanding, and multi-dimensional evaluation.

The Govern function emphasizes the importance of leadership, accountability structures, and clearly defined risk tolerances. Research in organizational governance consistently highlights that artificial intelligence initiatives fail not because the technology is flawed, but because institutional oversight is incomplete or reactive. The Map function corresponds with transparency and system characterization research. It requires organizations to identify the artificial intelligence system's purpose, data sources, model architecture, dependencies, limitations, and operating environment. This aligns with studies demonstrating that risk increases dramatically when artificial intelligence systems are insufficiently documented or poorly understood.

The Measure function reflects the extensive academic work on artificial intelligence performance evaluation, including robustness testing, bias assessment, and reliability metrics. Scholars have shown that without systematic measurement, artificial intelligence systems may gradually degrade in accuracy or become brittle under novel inputs. Finally, the Manage function embodies risk response and mitigation practices grounded in traditional risk management and control theory. It involves developing treatment plans, monitoring mechanisms, and adaptation strategies that ensure artificial intelligence systems remain aligned with operational expectations.

Complementing the Core functions are the NIST Profiles and Implementation Examples. Profiles enable organizations to tailor the framework to their specific contexts, consistent with research emphasizing that artificial intelligence risk is highly domain dependent. Implementation Examples offer practical guidance, supporting the translation of conceptual risk principles into concrete actions. For CISOs and cybersecurity leaders, these components supply a structured method for evaluating artificial intelligence applications such as phishing detection, anomaly analysis, automated triage, and threat intelligence processing.

While the NIST Artificial Intelligence Risk Management Framework provides a risk focused methodology for understanding and addressing artificial intelligence risks, ISO and IEC 42001 offers a systemic management system-based approach that institutionalizes artificial intelligence governance throughout the organization. This international standard draws from the extensive body of research supporting management system standards such as ISO 9001 and ISO 27001, which have long demonstrated the value of consistent processes, documented procedures, and continuous improvement cycles. ISO and IEC 42001 applies these principles specifically to artificial intelligence, recognizing that artificial intelligence governance requires coordinated planning, leadership engagement, operational controls, performance evaluation, and organizational learning.

Built around the Plan, Do, Check, Act improvement model, ISO and IEC 42001 provides a framework for integrating artificial intelligence governance into existing corporate processes. The planning stage aligns with academic recommendations that artificial intelligence initiatives begin with formalized objectives, ethical considerations, and a clear understanding of organizational context. The operational stage reflects research showing that consistent procedures for model development, data governance, validation, deployment, and human oversight significantly reduce the likelihood of emergent failures. The checking stage incorporates internal audits and performance evaluations, which are broadly supported in the literature as essential for maintaining trust and accountability in artificial intelligence systems. The acting stage ensures that evaluation results lead to organizational improvements, embodying the continuous learning principles emphasized in sociotechnical systems research.

For cybersecurity operations, ISO and IEC 42001 provide the governance infrastructure needed to support artificial intelligence driven tools such as automated threat detection, artificial intelligence assisted response platforms, and predictive analytics. The standard ensures that artificial intelligence risk mitigation strategies informed by the NIST framework do not remain conceptual but are operationalized into documented controls, training programs, and formal oversight activities. This relationship mirrors academic frameworks advocating that artificial intelligence risk management must be paired with institutional governance to achieve long term reliability and trustworthiness.

The complementarity of NIST AI RMF and ISO and IEC 42001 reinforce a central principle in artificial intelligence governance research: no single framework is sufficient for the multidimensional challenges artificial intelligence presents. NIST offers the analytic and evaluative tools needed to characterize and mitigate risk. ISO and IEC 42001 provide the system level structure that embeds those tools into ongoing operations. Consequently, an assessment conducted under the NIST frameworks such as evaluating the risk of model drift in an intrusion detection model can be operationalized through ISO and IEC 42001 requirements for performance monitoring, incident response, and continuous improvement.

By integrating these frameworks, organizations align themselves with leading research, regulatory expectations, and internationally recognized best practices. This alignment increases transparency, strengthens internal accountability, and enhances external trust. More importantly, it transforms artificial intelligence governance from a reactive series of controls into a proactive and resilient system capable of supporting secure, ethical, and effective artificial intelligence deployment across cybersecurity environments.

Defining Clear Decision Rights and Accountability

A foundational component of any mature artificial intelligence governance program is the explicit definition of decision rights and the unambiguous assignment of accountability. Research across organizational governance, complex systems, and risk management consistently reinforces that the absence of clearly delineated authority structures increases the likelihood of project failure, decision paralysis, and uncontrolled risk escalation. Within artificial intelligence governance, these concerns are magnified due to the adaptive behaviors, opaque decision pathways, and interdependent lifecycle stages inherent to artificial intelligence systems. As scholars have noted in studies examining sociotechnical systems, governance failures often emerge not from technical shortcomings but from unclear ownership and ambiguous authority. Thus, defining who is permitted to make which decisions, under what conditions, and with what evidence is not merely procedural; it is essential for safe and responsible artificial intelligence deployment.

Decision authority begins at the earliest stages of the artificial intelligence lifecycle, particularly during the evaluation and selection of potential use cases. Academic literature highlights that misalignment at this stage is a common cause of unsuccessful artificial intelligence initiatives, particularly when projects are pursued without strategic coherence or explicit evaluation of organizational risk tolerance. Determining whether an artificial intelligence powered threat intelligence platform should be pursued, for example, requires informed judgment about anticipated benefits, operational risk, and data sensitivity. Best practice research suggests that such decisions are most effective when made by cross functional leadership bodies such as an artificial intelligence governance committee or a strategic technology oversight board. These bodies typically include representatives from cybersecurity, information technology, legal, compliance, risk management, and domain specific business units. Their collective expertise enables a comprehensive evaluation of proposals, ensuring alignment with business strategy, regulatory requirements, and organizational values.

After use case approval, decision rights extend to model selection and validation. Research in machine learning governance underscores the importance of jointly evaluating model performance, interpretability, and computational feasibility. Selecting an algorithm for a fraud detection system, for example, requires balancing accuracy with transparency, fairness, and operational efficiency. In most organizations, technical teams such as data scientists, machine learning engineers, and security architects exercise primary authority in recommending models based on empirical evaluation. However, academic frameworks for risk governance emphasize that final approval for high impact artificial intelligence systems should rest with senior leadership, including the CISO or executive committees. This tiered decision structure ensures that technical expertise informs model selection while maintaining alignment with enterprise level risk tolerance.

One of the most sensitive decision domains concerns the acceptance of residual artificial intelligence risk. Extensive governance literature stresses that residual risk cannot be eliminated and must instead be formally acknowledged by individuals with appropriate authority. In artificial intelligence contexts, residual risks may include interpretability limitations, potential model drift, or incomplete mitigation of bias. Organizational research demonstrates that distributing risk acceptance across hierarchical levels according to magnitude is a best practice. For example, residual risk related to an internal document classification model may be accepted by a department leader,

50

whereas risks tied to customer facing artificial intelligence systems or critical security models typically require approval from the CISO, Chief Risk Officer, or Board of Directors. Formal risk acceptance documentation, which outlines identified risks, mitigation strategies, and justification for acceptance, is widely regarded as a mechanism for transparency and ethical accountability.

Decision authority must also be explicitly defined for artificial intelligence related incident response. Studies on high-risk automation systems indicate that delayed or uncertain leadership during incidents significantly increases operational and reputational harm. When an artificial intelligence model produces anomalous outputs, demonstrates drift, or is implicated in a security compromise, the organization must know exactly who is empowered to isolate the system, revert to a prior model version, initiate a forensic investigation, or notify regulators. Typically, the incident commander within the Security Operations Center or Computer Security Incident Response Team exercises operational control, supported by artificial intelligence technical leads and legal counsel. Academic recommendations stress the importance of predefined escalation pathways and explicit authority assignments to prevent hesitation during time sensitive events.

Parallel to the establishment of decision rights is the equally critical task of clarifying accountability. Academic research on governance consistently highlights that accountability structures must be explicit, enforceable, and traceable across the entire system lifecycle. In artificial intelligence contexts, accountability spans from initial system design to decommissioning, encompassing data stewardship, model development, deployment, monitoring, and ethical oversight.

Executive leadership, including the Chief Executive Officer or a Chief Artificial Intelligence Officer, carries overarching accountability for the governance of artificial intelligence within the enterprise. They ensure the existence, appropriateness, and effectiveness of the organizational artificial intelligence governance program. The CISO, as a central figure in cybersecurity governance, holds accountability for ensuring that artificial intelligence systems do not introduce unacceptable vulnerabilities and that artificial intelligence technologies used in security operations are themselves safeguarded. Academic research on cybersecurity governance strongly supports this role alignment.

Accountability then cascades to operational and technical teams. Data scientists and artificial intelligence engineers are accountable for adherence to ethical and technical standards during model development, including representative training data, robust validation, documentation, and bias mitigation. If biased outcomes occur due to insufficient evaluation or unaddressed training data issues, these teams are accountable for the oversight. Information technology and cybersecurity operations teams bear accountability for secure deployment and continuous monitoring. Empirical research on artificial intelligence system failures shows that many incidents stem from lapses in operational controls or insufficient monitoring, reinforcing the need for strong accountability for system upkeep.

Business unit leaders remain accountable for ensuring that artificial intelligence systems within their domain comply with policy, produce intended outcomes, and do not harm stakeholders. Legal and compliance teams hold accountability for ensuring adherence to regulations, privacy expectations, and ethical frameworks. Their role reflects findings from legal scholarship emphasizing that artificial intelligence governance must incorporate legal interpretability and regulatory foresight.

To operationalize accountability, research strongly recommends maintaining a comprehensive artificial intelligence system inventory, establishing clearly defined service level or operating level agreements, conducting regular audits, and fostering a culture that prioritizes transparency and learning. Scholars emphasize that punitive cultures undermine responsible artificial intelligence, whereas constructive accountability cultures promote continuous improvement and responsible innovation.

In summary, defining clear decision rights and establishing robust accountability mechanisms are fundamental pillars of mature artificial intelligence governance. They ensure responsible oversight, ethical alignment, and operational resilience. Without them, organizations risk encountering ambiguous responsibility, unmitigated risks, and preventable failures that undermine trust, compliance, and strategic outcomes.

Integrating AI Governance with Existing Risk Management Processes

The accelerating adoption of artificial intelligence across modern enterprises has introduced profound opportunities, but it has also expanded the risk surface in ways that organizations are still learning to navigate. Effective governance cannot rely solely on isolated artificial intelligence oversight mechanisms; it requires embedding artificial intelligence governance directly into the organization's established enterprise risk management framework. This alignment is not a peripheral adjustment or a procedural enhancement. Rather, it is a strategic necessity to ensure that artificial intelligence is treated not as a novelty or a siloed technological domain, but as a core component of the organization's overall risk posture. Integrating artificial intelligence governance principles into enterprise risk management enables a unified, coherent, and enterprise wide approach to identifying, assessing, mitigating, and monitoring risks. Without this integration, organizations risk developing fragmented oversight structures that obscure interdependencies, create blind spots, and weaken the organization's capacity to respond to emerging artificial intelligence driven threats.

The first step in this integration is the deliberate inclusion of artificial intelligence specific risks into the organization's existing enterprise risk taxonomy. Most organizations already categorize risks across domains such as operational, financial, strategic, compliance, and cybersecurity. Artificial intelligence introduces novel risks that intersect with each of these domains. Data related risks such as privacy violations, bias, and data integrity issues; model related risks including drift, opacity, and susceptibility to adversarial attacks; ethical risks involving fairness, accountability, and transparency; and operational risks such as system failures or unintended consequences must be formally incorporated into the risk catalog. Mapping these artificial intelligence risks to existing categories preserves continuity with established reporting mechanisms while acknowledging their distinct characteristics. For example, the risk of discriminatory outcomes from an artificial intelligence model may be associated with compliance, ethical, and reputational concerns, while the risk of adversarial manipulation aligns with cybersecurity risk. This mapping preserves the familiarity of existing frameworks while expanding them to accommodate the complexities of artificial intelligence.

Risk identification processes must also evolve to capture the full spectrum of artificial intelligence-related risks. Traditional methodologies such as risk workshops, surveys, threat modeling, and incident analysis must be adapted to explicitly examine artificial intelligence systems. During cybersecurity assessments, analysts must consider not only traditional network vulnerabilities but also the potential for attacks that exploit weaknesses in artificial intelligence models. Operational risk assessments must evaluate the impact of artificial intelligence system failures or performance degradation on business continuity. Including subject matter experts such as data scientists, artificial intelligence engineers, ethicists, privacy professionals, and business stakeholders ensures that artificial intelligence risks are surfaced, articulated, and understood comprehensively. This cross-disciplinary participation reinforces the shared accountability necessary for mature artificial intelligence governance.

Once cataloged, artificial intelligence risks must be evaluated through the same assessment frameworks used for enterprise risks, incorporating both likelihood and impact. Quantifying artificial intelligence risk is sometimes more complex due to the adaptive nature of artificial intelligence, the opacity of certain model classes, and the ethical dimensions associated with automated decision making. Nevertheless, enterprise risk management methodologies can accommodate both qualitative and quantitative approaches. Assessments should consider financial implications, regulatory exposure, reputational impact, operational disruption, and erosion of stakeholder trust. Organizations may need to develop tailored impact scales or scoring models for artificial intelligence risks, particularly for evaluating harms related to bias or the cascading effects of model drift. The goal is to ensure comparability across risk types so that artificial intelligence risks can be prioritized appropriately within the enterprise risk portfolio.

Artificial intelligence governance must also align with the organization's risk appetite and tolerance statements. Enterprise risk management defines how much risk the organization is willing to assume in pursuit of its objectives. Artificial intelligence initiatives especially those involving sensitive data, automated decision systems, or mission critical operations must be evaluated against these thresholds. For example, in a low tolerance environment for regulatory noncompliance, an artificial intelligence initiative involving large scale processing of personal data may require enhanced scrutiny, heightened controls, or formal rejection. Enterprise risk committees must therefore play a central role in approving and regularly reviewing artificial intelligence initiatives. This oversight ensures that adoption is not only innovative but disciplined, predictable, and aligned with enterprise level risk boundaries.

Risk treatment strategies must be managed holistically. Instead of creating a separate track for artificial intelligence risk treatment, mitigation strategies should be integrated into existing operational, cybersecurity, and compliance programs. Controls such as data validation routines, bias detection tools, secure training practices, and automated drift monitoring can be embedded into existing governance structures. This ensures clear ownership, efficient resource allocation, and alignment with policies already in effect. Moreover, enterprise risk management's existing treatment strategies including avoidance, reduction, sharing, and acceptance apply equally to artificial intelligence risks. Harmonizing these approaches reinforces organizational coherence and reduces redundant or conflicting efforts.

Continuous monitoring remains essential. Artificial intelligence systems require ongoing oversight to detect model drift, data quality issues, adversarial probing, and ethical compliance failures. Integrating artificial intelligence specific risk indicators into enterprise dashboards and reporting mechanisms ensures visibility for leadership and the board. Examples include drift detection rates, bias scores, attempts to attack artificial intelligence models, performance degradation metrics, and policy compliance indicators. By embedding these indicators into enterprise reporting, organizations maintain a unified risk narrative and can rapidly respond to emerging threats.

One of the greatest strategic benefits of integrating artificial intelligence governance with enterprise risk management is the elimination of siloed oversight structures. Standalone artificial intelligence governance programs, although well intentioned, often lack alignment with broader organizational priorities and can create fragmentation. Embedding artificial intelligence governance within enterprise risk management ensures that artificial intelligence is evaluated within the context of strategic objectives, interdependent business processes, and cross functional workflows. This holistic lens is essential for uncovering systemic or cascading risks that may not be apparent when artificial intelligence is viewed in isolation. Enterprise risk management's inherent cross functional model promotes collaboration between cybersecurity, information technology, legal, compliance, and business units, enabling a comprehensive approach to multifaceted artificial intelligence risks.

Integration also enhances operational efficiency. Organizations often invest significant resources into enterprise risk management systems, personnel, and processes. Leveraging these existing capabilities avoids the cost, redundancy, and confusion associated with creating parallel governance structures. Modern enterprise risk platforms can be adapted to include artificial intelligence risk registers, control libraries, incident workflows, and monitoring dashboards. This consolidation strengthens risk visibility, reduces administrative burden, and supports a unified governance culture.

A mature integration model may include well defined roles for enterprise risk management committees, supported by specialized artificial intelligence risk working groups. These working groups supply technical expertise and conduct deep analysis while keeping their findings aligned with enterprise level risk priorities. Technology plays a supporting role, enabling enterprise risk systems to host artificial intelligence specific assessments, workflows, and controls.

Ultimately, successful integration depends on cultivating a risk aware culture that extends to artificial intelligence. Communication, education, and ongoing awareness efforts are essential to ensure that employees understand the importance of artificial intelligence governance and their role in managing associated risks. When artificial intelligence governance is embedded within enterprise risk management, organizations can fully leverage artificial intelligence as a driver of innovation while maintaining a resilient and defensible risk posture. This approach transforms artificial intelligence governance from a specialized discipline into a foundational element of enterprise resilience and strategic execution.

Conclusion: Governance as the Operating System for AI in Security

AI governance is the discipline that turns powerful capability into controlled advantage. This chapter has established that as AI becomes embedded in cybersecurity operations and enterprise decision systems, governance cannot remain a policy exercise or a standalone program. Governance must function as an interconnected operating model that defines intent, assigns decision rights, enforces accountability, and sustains oversight across the full AI lifecycle. Without this structure, AI introduces unmanaged risk through opaque behavior, adversarial exposure, performance decay, and ethical drift. With this structure, AI becomes a secure, auditable, and strategically aligned contributor to organizational resilience.

The AI Governance Lifecycle Model reinforced an essential leadership truth. Trustworthy AI is not achieved at deployment. Trustworthy AI is built through disciplined phases that begin with clear problem definition, continue through data governance, validation, risk and compliance assessment, and extend into operational controls, monitoring, and responsible retirement. Each phase exists because AI behaves differently than traditional software. AI can drift, learn unstable correlations, absorb bias from data, and become a target for manipulation. Governance therefore must be continuous, evidence-based, and designed for an environment where both threats and models evolve.

This chapter also clarified that governance works only when accountability becomes explicit and enforceable. Steering committees, governance leads, technical owners, legal and compliance partners, and operational teams must share a common framework of responsibility, supported by clear escalation paths and documented decision authority. Decision rights must define who approves high-impact use cases, who accepts residual risk, who can pause or roll back models during incidents, and who is accountable when automation creates harm. These structures reduce ambiguity, enable faster response under pressure, and prevent diffusion of responsibility across technical and business lines.

Finally, durable AI governance cannot be siloed. Integrating AI governance into enterprise risk management ensures AI risks are visible, categorized, prioritized, and monitored alongside other strategic exposures. This integration aligns AI initiatives with risk appetite, reinforces consistent treatment strategies, and enables board-level reporting using language and mechanisms leadership already trusts. When AI governance merges with ERM, AI adoption shifts from novelty to discipline, and from isolated tooling to enterprise posture.

The leadership imperative is simple. Security executives must treat governance as a prerequisite for scale. AI can accelerate detection and response, but only governance provides the guardrails that keep automation defensible, ethical, and resilient. Organizations that build this governance architecture will move faster with confidence. Organizations that do not will move quickly into a future where critical security decisions are made by systems they cannot fully explain, control, or justify.

Chapter 3: Ethical Considerations and Responsible AI Deployment

Artificial intelligence is rapidly reshaping the security landscape, offering unprecedented speed, scale, and analytical depth, yet its growing influence introduces profound ethical, operational, and governance challenges that security leaders can no longer afford to ignore. As AI systems begin to detect threats, shape decisions, and act on behalf of the enterprise, CISOs and their teams must confront a new reality where technical effectiveness is inseparable from ethical responsibility. This chapter explores the critical dimensions of responsible AI deployment in cybersecurity, including the ethical imperatives that guide AI-driven defense, the identification and mitigation of bias within security tools, the vital role of human oversight in maintaining accountability, and the development of an Ethical AI Charter that anchors governance in values, fairness, and organizational integrity. Together, these elements form the foundation of a security program that leverages AI as a powerful ally without compromising the trust, transparency, or ethical principles that define resilient and responsible cyber defense.

The Ethical Imperative in AI-Driven Security

The integration of artificial intelligence into cybersecurity operations, while promising enhanced defense capabilities, simultaneously surfaces a profound ethical imperative. This imperative stems from the inherent nature of AI, its reliance on data, its potential for complex decision-making, and its growing autonomy. When AI systems are tasked with protecting sensitive information, critical infrastructure, and individual privacy, the ethical considerations surrounding their deployment become paramount. The goal is not merely to build more robust defenses, but to ensure these defenses are fair, equitable, transparent, accountable, and ultimately aligned with humanity's best interests, particularly in the high-stakes domain of security. Ethical AI security requires solutions that not only perform effectively but also operate in ways that can be clearly understood, validated, and trusted.

One of the most significant ethical challenges lies in the area of fairness and bias. AI models learn from the data they are trained on. If this data reflects existing societal biases, the AI will inevitably learn and perpetuate these biases, often in ways that are subtle and difficult to detect. In cybersecurity, this can manifest in several concerning ways. For example, an AI-powered threat detection system trained on data that underrepresents certain demographic groups might be less effective at identifying threats originating from or targeting those groups. This could lead to disparate levels of protection, effectively creating digital blind spots for marginalized communities. Imagine an AI anomaly detection system used to identify fraudulent activity that is trained on historical transaction data in which certain groups were disproportionately flagged due to socioeconomic factors unrelated to actual fraud risk. The AI could continue to unfairly flag legitimate transactions from individuals within those groups, leading to unnecessary scrutiny, account restrictions, reduced access, and an erosion of trust. The consequence is not just a technical failure but a social and ethical one, where the very tools designed to protect end up discriminating.

The challenge of bias in AI security is not confined to detection systems. It can also appear in automated response mechanisms. If an AI is tasked with responding to a security incident and its decision-making is biased, it could inadvertently escalate certain situations or prioritize responses based on flawed criteria. An AI system designed to manage access control might grant privileged access to individuals who pose a higher risk simply because they belong to a group historically favored in the training data.

Conversely, it might deny access to legitimate users unfairly associated with higher risk profiles. The absence of human oversight in such automated responses exacerbates the problem, as the AI's decisions can be swift and far-reaching, leaving little room for immediate correction. Ensuring fairness requires a proactive and continuous effort to identify and mitigate bias in data, algorithms, and outputs.

Transparency and explainability present another critical ethical hurdle. Many advanced AI models, particularly deep learning systems, operate with internal logic so complex that even their creators struggle to fully understand how decisions are made. In cybersecurity, where accountability is paramount, this opacity is deeply problematic. When an AI system flags a user as malicious, blocks a transaction, or initiates an automated defensive action, it is essential to understand why. Without transparency, it becomes impossible to validate decisions, debug or improve systems, assign responsibility when harm occurs, or satisfy regulatory requirements that demand explainable outcomes. Consider an AI used in network intrusion detection. If it incorrectly identifies legitimate activity as malicious and triggers a system lockdown, the security team needs to know precisely what caused the false positive. Was it an unusual packet signature misclassified by the model, or was it a legitimate pattern associated with a unique business process? Without explainability, teams are left guessing, wasting time and resources, or failing to identify the true cause. This lack of transparency undermines confidence in AI-driven tools and can slow or prevent adoption. Advancements in explainable AI aim to address these concerns by enabling greater interpretability and traceability of decisions.

The potential for AI systems to inadvertently cause harm or discrimination is a broad ethical concern that permeates many aspects of AI deployment in security. Beyond bias in data, unintended consequences can arise from the complex interactions between autonomous systems and dynamic operational environments. For example, an AI-powered autonomous cyber defense system might be programmed to neutralize threats aggressively. In a complex network environment, its rapid and rigid actions could disrupt critical business operations, cause financial damage, or even endanger lives if the system controls essential infrastructure such as electrical grids or medical equipment. The mindset of moving fast and accepting breakage, common in experimental technology development, is wholly unacceptable in environments where stability, reliability, and safety are paramount.

A more everyday example is an AI system designed to manage phishing responses. If the system automatically flags and quarantines emails based on sophisticated pattern recognition, it might inadvertently block important communications such as financial alerts, legal notifications, medical reminders, or job offers. The harm caused by such misclassification could be substantial, especially if the error persists unnoticed. Additionally, the use of AI in surveillance and monitoring introduces significant privacy concerns. While these tools may be justified for threat detection, their intrusive nature can infringe upon individual rights if not governed with clear limitations and strong protections. The ethical imperative is to ensure that the pursuit of security does not violate privacy, dignity, or the fundamental right to fair treatment. Achieving this balance requires rigorous impact assessments, clear data usage guidelines, and robust mechanisms for oversight and redress.

Accountability poses another difficult challenge. When an AI system errs, who is responsible? The developers who created the algorithm? The organization that deployed it? The data scientists who trained it? The security analysts who relied on its outputs? Traditional accountability frameworks rely on concepts of human intent and agency, which are difficult to apply when AI systems operate autonomously or semi-autonomously. If an AI-driven security tool misclassifies traffic, unintentionally causes downtime, or fails to detect a novel attack vector, determining liability becomes a complex ethical and legal puzzle. This ambiguity creates responsibility gaps that undermine trust and leave affected parties without clear recourse. Ethical AI deployment requires proactive guidance on roles, responsibilities, liability, incident response expectations, and oversight from initial design through ongoing use.

Adding to these challenges is the increasing sophistication of AI in offensive cybersecurity. AI-powered malware, automated exploitation tools, and hyper-targeted social engineering campaigns elevate the threat landscape dramatically. While defenders rely on AI to detect such threats, adversaries use the same capabilities for malicious purposes. This creates a dual-use dilemma. Organizations developing AI systems must consider how their tools could be misused. Ethical development demands safeguards such as strict access controls, secure development pipelines, and deliberate restraint regarding the release or proliferation of powerful AI capabilities that could be weaponized.

The ethical imperative in AI-driven security is not a static checklist but an evolving challenge requiring continuous dialogue among technologists, ethicists, policymakers, and the public. Trust in AI security solutions will depend on a demonstrated commitment to ethical principles throughout the AI lifecycle, from conceptual design to deployment and eventual decommissioning. It requires cultivating a culture of ethical awareness among AI practitioners, encouraging them to reflect on the implications of their work, anticipate unintended consequences, and prioritize human well-being. Ultimately, the responsible deployment of AI in cybersecurity is not only a matter of technical expertise. It is a moral obligation to ensure that AI is used to enhance security for all people without compromising fairness, privacy, or fundamental human rights. The future of secure digital environments depends on our ability to navigate these ethical complexities with diligence, foresight, and an unwavering commitment to human values.

Vignette: When the AI Got It Wrong

The security operations center at Meridian Health Services was known for its efficiency. With thousands of employees, a sprawling digital footprint, and sensitive patient data spread across interconnected systems, the organization had invested heavily in AI-driven cybersecurity tools. The crown jewel of this investment was Vigilant Shield, an autonomous threat detection and response platform that analyzed network behavior in real time and initiated defensive actions without human intervention. The system had already prevented several ransomware attempts and was hailed internally as a breakthrough in operational security.

On a quiet Thursday morning, however, Vigilant Shield triggered an automated lockdown of several clinical systems, including the scheduling interface used by physicians and nurses. The system flagged unusual login behavior from the account of an oncology nurse named Carla, interpreting the activity as an early-stage credential compromise. Without waiting for human confirmation, Vigilant Shield revoked her access, isolated her workstation, and blocked all outbound communications associated with her credentials. Within minutes, patient appointments were being delayed, electronic records were temporarily inaccessible, and clinical staff were forced to revert to manual processes.

When the security team began investigating, they discovered that Carla had been covering for a colleague in a different wing of the hospital and had logged in from an unfamiliar workstation. The unfamiliar geolocation pattern, combined with an unusual series of rapid chart accesses, triggered the AI's internal risk threshold. While technically consistent with the model's learned patterns, the interpretation was entirely wrong. Carla was doing her job, not compromising the network.

As the investigation continued, another troubling discovery emerged. The model's training data had disproportionately flagged login anomalies from night-shift nurses in specific departments, many of whom happened to share similar demographic profiles. The AI had internalized subtle correlations between race, age, and work patterns without anyone realizing it. In Carla's case, these hidden correlations contributed to the elevated threat score that triggered the automatic lockdown.

The consequences rippled across the organization. Clinical operations were disrupted for nearly two hours, several patient treatments were delayed, and trust in the new security system was shaken. Even more concerning was the realization that the AI's decision-making process could not be fully explained. The model offered risk scores and confidence levels but provided no meaningful interpretation of which factors mattered most. As the security team struggled to understand how the false positive occurred, it became clear that they were operating inside a black box.

Leadership convened an emergency meeting. The CIO demanded to know why the system had acted without human oversight. The Chief Privacy Officer questioned whether the AI's underlying bias posed legal risks. The Chief Medical Officer was concerned about patient safety. The CISO found himself at the center of a complex ethical and operational dilemma. The system had been deployed to strengthen security, yet its opaque and biased behavior had undermined both safety and trust.

The incident became a turning point. Meridian implemented a new governance framework that required human review for all high-impact automated actions, mandated explainability in future AI acquisitions, and established a cross-functional oversight committee to evaluate ethical risks. The hospital also retrained Vigilant Shield using more diverse and representative data, added contextual features to reduce overreliance on demographic proxies, and embedded privacy safeguards to limit the unnecessary collection of sensitive information.

The event served as a sobering reminder that even well-intentioned AI systems can cause harm when oversight, transparency, and fairness are not treated as core operational controls. For Meridian, the lesson was clear. Strengthening security must never come at the expense of the people the system is meant to protect.

Identifying and Mitigating AI Bias in Security Tools

The pervasive nature of bias in artificial intelligence systems deployed within the cybersecurity domain presents a significant ethical and operational challenge. When AI models are entrusted with the critical tasks of threat detection, anomaly identification, or resource allocation, any inherent bias can lead to profoundly inequitable and ineffective outcomes. These biases, often ingrained within the very data used to train these sophisticated algorithms, can manifest as flawed threat assessments, discriminatory access controls, or disproportionate impacts on certain user groups. The consequence is a cybersecurity posture that is not only less secure but also ethically compromised, failing to uphold principles of fairness and equity. Therefore, a proactive and systematic approach to identifying and mitigating AI bias is no longer an option but a fundamental necessity for responsible AI deployment in security.

Identifying the Roots of Bias in Security AI

The journey toward unbiased AI in security begins with a deep understanding of where bias originates. The primary culprits are invariably the datasets and the algorithms themselves, each presenting unique challenges that demand careful scrutiny.

Data-Centric Bias: The bedrock of any AI model is its training data. If this data is unrepresentative, incomplete, or reflects historical societal prejudices, the AI will inevitably absorb and amplify these flaws. In cybersecurity, this can take several forms:

Underrepresentation of Specific Threat Vectors or User Groups: Imagine an AI threat detection system trained primarily on data from Western markets. It might be significantly less adept at identifying sophisticated attack patterns originating from regions with distinct cybercriminal methodologies or infrastructure. Similarly, if the training data for an anomaly detection system lacks sufficient examples of legitimate user behavior from certain demographic groups or individuals with unique work patterns, it could unfairly flag their normal activities as suspicious. This could lead to legitimate users being subjected to unnecessary security checks or even having their access revoked, impacting productivity and eroding trust.

Historical Biases in Incident Data: Cybersecurity incident logs often reflect past security priorities and enforcement actions. If, for instance, historical data shows a disproportionate number of alerts or investigations focused on specific types of users or systems due to organizational biases at the time, an AI trained on this data might continue to prioritize threats from these same sources, even if current threat landscapes have shifted. For example, an AI trained on early cybersecurity data might have been overly focused on traditional network intrusions, neglecting the evolving landscape of insider threats or sophisticated phishing campaigns that target human vulnerabilities, regardless of the user's technical proficiency.

Labeling Errors and Subjectivity: The process of labeling data – identifying what constitutes a threat, an anomaly, or normal behavior – is often performed by human annotators. Human subjectivity, unconscious biases, and varying levels of expertise can introduce errors and inconsistencies into these labels. If, for instance, data annotators have a preconceived notion about the types of users who are more likely to engage in malicious activity, their labeling of training examples could inadvertently skew the AI's learning process. This is particularly problematic in areas like user behavior analytics, where distinguishing between risky behavior and benign deviations can be highly subjective.

Data Drift and Concept Drift: The cybersecurity landscape is in a constant state of flux. New attack techniques emerge daily, legitimate user behaviors evolve, and the underlying infrastructure changes. If the training data used for an AI model is not periodically updated to reflect these changes (data drift), or if the very definition of what constitutes a threat or anomaly shifts (concept drift), the model's performance will degrade, and its decisions can become biased against current realities. An AI trained on pre-pandemic network traffic patterns, for example, might struggle to accurately assess the security of a predominantly remote workforce without retraining.

Algorithmic Bias: While data is often the primary source, the algorithms themselves can also introduce or amplify bias:

Model Complexity and Opacity: Highly complex models, such as deep neural networks, can sometimes learn spurious correlations that are not genuinely indicative of the underlying phenomena they are meant to model. These complex models, often referred to as "black boxes," can be difficult to interpret, making it challenging to identify precisely why a particular decision was made. If a model learns to associate certain patterns with malicious activity that are coincidentally present in data from a specific demographic, it can perpetuate bias without clear intent.

Feature Selection and Engineering: The process of selecting and transforming raw data into features that an AI model can understand can also introduce bias. If certain features are inadvertently chosen or engineered in a way that disproportionately represents or penalizes specific groups, this bias will be carried into the model. For instance, if an AI system for predicting security vulnerabilities relies heavily on features related to the age of software components without considering their actual risk profile, older but well-maintained systems might be flagged unfairly.

Optimization Objectives: AI models are trained to optimize specific objectives, such as minimizing error rates. If these objectives are not carefully defined to include fairness metrics alongside accuracy, the model might achieve high accuracy by performing exceptionally well on the majority group while performing poorly on minority groups. This can lead to a statistically accurate but ethically unacceptable outcome. For example, an intrusion detection system might achieve a high overall accuracy by being very good at detecting threats on standard enterprise networks, but if it consistently misses novel attack vectors that are more prevalent in specialized research environments, it exhibits a performance bias.

Mitigation Strategies: Building a More Equitable Security AI

Once the potential sources of bias are understood, a multi-faceted approach is required to actively mitigate them. This involves a combination of pre-processing data, refining algorithms, and implementing rigorous testing and monitoring frameworks.

1. Data Augmentation and Rebalancing:

Addressing underrepresentation in datasets is a crucial first step. Data augmentation techniques involve artificially expanding the dataset by creating modified copies of existing data. For cybersecurity, this could include:

Synthetic Data Generation: Using generative adversarial networks (GANs) or other synthetic data generation methods to create realistic but artificial examples of threat events or user behaviors that are underrepresented in the original dataset. This can help to balance the dataset and expose the AI to a wider range of scenarios. For example, if an AI is being trained to detect zero-day exploits, and real-world examples are scarce, synthetic data can simulate such exploits based on known attack characteristics.

Oversampling and Undersampling: If certain classes of data are underrepresented, oversampling involves duplicating instances from those classes. Conversely, undersampling involves removing instances from overrepresented classes. These techniques aim to create a more balanced distribution of data, preventing the AI from being overly influenced by the majority class.

Perturbation Techniques: Modifying existing data points slightly by adding noise or making minor alterations can create new variations that increase the dataset's diversity. This can be particularly useful for text-based data, like phishing email content, where subtle word changes or rephrasing can generate new training examples.

2. Algorithmic Debiasing Techniques:

Beyond data manipulation, algorithms can be designed or modified to actively reduce bias:

Adversarial Debiasing: This sophisticated technique involves training two competing neural networks: one that tries to predict the target variable (e.g., whether an event is malicious) and another that tries to predict a sensitive attribute (e.g., user demographic). The primary network is trained to perform its task accurately while simultaneously confusing the adversary network, thus discouraging it from learning to predict the sensitive attribute based on the features. This forces the model to learn representations that are predictive of the target without relying on biased correlations with sensitive attributes.

Regularization Techniques: Incorporating regularization terms into the AI model's loss function can penalize reliance on biased features. This encourages the model to find solutions that are less sensitive to attributes that might be correlated with protected characteristics.

Fairness-Aware Machine Learning Algorithms: Researchers are developing algorithms specifically designed to incorporate fairness constraints into their learning processes. These algorithms aim to optimize for both predictive accuracy and various fairness metrics simultaneously. Examples include algorithms that enforce demographic parity (ensuring equal prediction rates across groups) or equalized odds (ensuring equal true positive and false positive rates across groups).

3. Rigorous Testing and Validation for Equity:

The development lifecycle must include robust testing phases specifically designed to uncover and quantify bias. This goes beyond standard accuracy metrics:

Disaggregated Performance Metrics: Instead of evaluating the AI's performance on the entire dataset, it is crucial to measure its performance across different subgroups. This means calculating metrics such as accuracy, precision, recall, and false positive rates for each demographic, geographic, or operational group. This granular analysis will reveal if the AI is disproportionately failing for certain segments of users or scenarios. For example, a threat detection system must be evaluated not just on its overall ability to catch malware, but on its ability to catch the same malware variants when deployed in different network environments or targeting different types of endpoints.

Bias Audits and Stress Testing: Regularly conduct formal bias audits where the AI system is subjected to controlled tests designed to expose its weaknesses regarding fairness. This involves creating specific test cases that are likely to trigger biased responses. Stress testing involves pushing the AI to its limits with edge cases and adversarial examples to see how its performance and fairness metrics hold up under pressure.

Red Teaming and Ethical Hacking: Employing internal or external "red teams" to actively try to circumvent or exploit the AI security system can reveal vulnerabilities and biases that might not be apparent through standard testing. These teams can simulate attacks that might specifically target perceived weaknesses in the AI's handling of certain user types or traffic patterns.

Human-in-the-Loop Review: For AI systems making critical decisions, implementing a "human-in-the-loop" mechanism is essential. This allows human analysts to review and override AI decisions, especially those that appear unusual or potentially biased. This provides a vital safeguard against AI errors and allows for continuous feedback on the AI's performance and fairness. The insights gained from these reviews can then be fed back into the AI's retraining process.

4. Continuous Monitoring and Feedback Loops:

Bias is not a static problem; it can emerge or evolve over time. Therefore, ongoing monitoring is critical:

Real-time Performance Monitoring: Deploying AI security tools requires continuous monitoring of their performance in the production environment. This includes tracking key metrics, but also actively looking for discrepancies in how the AI is performing across different user groups or operational contexts. Automated alerting systems can be set up to flag sudden drops in performance for specific subgroups.

Drift Detection: Implement mechanisms to detect data drift and concept drift. If the incoming data or the underlying concepts change significantly from the training data, the AI's performance may degrade, and biases can emerge. Prompt retraining or recalibration of the model is necessary in such cases.

Feedback Channels: Establish clear channels for users and security analysts to report suspected bias or unfair outcomes. These reports should be investigated thoroughly and used to improve the AI models. A system that allows a user to easily flag a security alert as "unfair" or "incorrect" provides invaluable qualitative data.

By adopting these strategies, organizations can transition from deploying AI tools simply for efficiency to implementing AI responsibly, with fairness and integrity at the core. Mitigating bias is not a narrow technical challenge but a broad ethical obligation. Building equitable, transparent, and accountable AI systems ensures that enhanced cybersecurity does not come at the cost of fairness, trust, or the rights of the individuals these systems are meant to protect.

Ensuring Transparency and Explainability in XAI

As artificial intelligence becomes foundational to enterprise security operations, the opacity of many advanced models presents a strategic and operational challenge that executives can no longer overlook. These systems often function as powerful yet opaque "black boxes," capable of producing accurate predictions while offering minimal insight into how those decisions were formed. In the security domain, where accountability, regulatory scrutiny, and high-stakes decision-making converge, lack of transparency is not merely a technical limitation. It is a material business risk. This reality elevates Explainable AI from a desirable enhancement to an indispensable requirement for any security program seeking trustworthiness, resilience, and executive-level assurance.

Transparency is indispensable for reliable incident response and operational decision-making. Security teams must understand why an AI flagged a specific behavior, escalated an alert, or classified an event as malicious. Without explainability, analysts are forced into a binary posture of either overreliance or skepticism, impairing the organization's ability to act decisively. Explainable AI provides analysts with the contextual intelligence needed to validate threats, reduce investigative time, and avoid unnecessary disruptions. A mature XAI capability does not simply highlight that a data exfiltration event is suspicious; it reveals the anomalous traffic pattern, the deviation from the user's historic baseline, the unusual destination address, and the contextual factors that distinguish the event from legitimate business processes. This turns opaque alerts into actionable, auditable intelligence.

Regulatory compliance further underscores the need for explainability. Data protection and AI governance frameworks increasingly demand that organizations justify automated decisions that influence user access, behavioral assessments, or incident escalation. Without clear explanations, organizations face compliance gaps, audit failures, and legal exposure. In industries such as finance, healthcare, government, and critical infrastructure, explainability is no longer optional. It is a formal requirement for demonstrating due diligence, maintaining operational integrity, and upholding public trust. XAI provides the mechanisms needed for traceability and accountability, enabling organizations to show not only what a model decided, but why it decided it.

Trust is the final, and perhaps most strategic, dimension of explainability. Executives must foster an environment in which employees, security teams, customers, and regulators trust the AI systems integrated into daily operations. When security decisions appear arbitrary or inexplicable, resistance grows, user adoption declines, and the credibility of the entire security program suffers. XAI transforms AI from an inscrutable authority into a transparent partner. By revealing the model's reasoning, the limits of its capabilities, and the confidence behind its assessments, XAI strengthens internal trust and empowers leaders to automate more confidently without compromising control or oversight.

Achieving high-quality explainability requires a multilayered architecture that integrates both interpretable model design and post-hoc explanation capabilities. Intrinsically interpretable models, such as decision trees, rule-based systems, and generalized linear models, offer clarity through their explicit logic. For straightforward security tasks such as rule enforcement or basic classification, these models reveal the exact conditions that triggered an alert. A decision tree can illustrate the precise sequence of conditions that labeled an email as malicious, while a linear model can highlight the relative influence of specific features in a risk score. However, modern cybersecurity demands extend far beyond the capacity of these simpler models. Detecting polymorphic malware, uncovering zero-day behaviors, or modeling complex insider risk patterns requires techniques far more sophisticated than what inherently transparent models can provide.

For these more complex systems, post-hoc interpretability becomes essential. Feature attribution methods such as SHAP values give executives and analysts the ability to understand the weighted contributions of each input variable to the model's output. SHAP can explain, with mathematical clarity, how the model transformed raw input into a final prediction. In a network threat scenario, SHAP may reveal that the decisive factors were the unusual port usage, suspicious destination addresses, deviations in traffic volume, and timing anomalies. LIME complements this with local approximations, delivering human-readable explanations for specific decisions and supporting rapid review of events that deviate from expected baselines.

In addition to attribution, example-based approaches deepen understanding by presenting the closest matches within the model's learned experience. When AI classifies a code artifact or a network packet as malicious, example-based XAI surfaces similar cases from training data, enabling analysts to compare current events with historically validated threats. Counterfactual explanations offer particularly valuable operational insight by describing exactly what minimal change would have transformed a flagged event into a non-alert state. In access control scenarios, a counterfactual might show that if a login attempt had originated from an approved subnet or occurred during a typical work window, the alert would not have triggered. This helps organizations refine both policy and model parameters.

XAI also leverages visualization methods that convert complex model behavior into accessible insights. Neural network visualizations can reveal which internal patterns correspond to specific attack signatures. High-dimensional clustering diagrams can illustrate how the model distinguishes normal behavior from anomalies. These tools are invaluable for security teams that must audit AI systems, understand evolving threat landscapes, and translate model logic into risk-based decisions.

The cybersecurity context introduces unique XAI considerations that executives must address proactively. First, explanations themselves must be secure. Adversaries may attempt to exploit explainability mechanisms to reverse-engineer model behavior or craft evasive attacks. XAI systems must therefore incorporate adversarial robustness and safeguards that prevent explanation leakage from becoming a strategic vulnerability. Second, explanations must be tailored to different stakeholders. Executives require high-level summaries aligned with risk posture, compliance teams require audit-ready justifications, and analysts require technical details for triage. Effective XAI systems offer multiple layers of abstraction to meet these distinct needs.

Integrating explainability into Security Operations Centers is now a strategic requirement for modern enterprises. Analysts should receive explanations alongside alerts, enabling immediate understanding of the factors that elevated risk and accelerating triage cycles. SIEM and SOAR platforms enhanced with XAI can contextualize alerts with top contributing features, similar historical incidents, correlated signals, and confidence measures, providing a comprehensive operational picture. This not only reduces the cognitive burden on analysts but strengthens governance and reduces incident response times.

Explainability also drives continuous model improvement. By analyzing explanations for misclassifications or unexpected behaviors, organizations can identify systemic flaws in model logic, dataset gaps, problematic features, or newly emerging behavioral patterns. This creates a powerful feedback loop that improves detection accuracy, enhances fairness, and strengthens resilience over time. XAI therefore becomes both a defensive capability and a strategic accelerator for AI maturity.

Ultimately, explainable AI is essential to responsible security modernization. It enables organizations to adopt advanced AI capabilities while maintaining clarity, accountability, and user trust. Executives who prioritize explainability create a security environment in which AI is not only powerful but also predictable, auditable, and aligned with organizational values. In a threat landscape defined by complexity and speed, explainability transforms AI from a black box to a strategic asset, ensuring that security decisions are both intelligent and justifiable. It reinforces governance, enhances operational resilience, and enables organizations to scale AI confidently and responsibly across the enterprise.

Human Oversight and Accountability in AI Systems

While artificial intelligence significantly transforms cybersecurity operations by augmenting detection, analysis, and response, its deployment introduces a corresponding need for intentional, disciplined human oversight. Cybersecurity remains an inherently complex and dynamic field, shaped by shifting threat actors, evolving attack methods, and continuous technological change. Automated systems, even highly advanced ones, operate within the constraints of their training data, their design assumptions, and the limits of machine logic. They cannot account for organizational culture, shifting business priorities, ethical obligations, or the broader consequences of an incorrectly automated decision. For this reason, the integration of AI into security operations must be grounded in the unwavering principle that humans remain the final arbiters of security outcomes. AI enhances capabilities, but it is human judgment that ensures those capabilities are used responsibly.

This necessity becomes especially clear when considering the human-in-the-loop model, which represents the most direct form of human oversight. In this model, AI conducts analysis, identifies anomalies, and generates recommendations, yet the final decision rests firmly with a human analyst. This approach is critical in environments where security decisions can have far-reaching implications, such as disrupting core operational systems, locking out essential personnel, initiating forensic-level investigations, or triggering compliance reporting. The analyst must evaluate not only the indicators AI surfaces but also contextual factors AI cannot fully interpret, such as business cycles, urgent projects, known behavioral variations among user groups, executive travel patterns, seasonal workload changes, or nuanced exceptions that fall outside AI's learned boundaries. The human-in-the-loop model ensures that sensitive decisions reflect organizational values, operational realities, and ethical considerations. However, this model also places a significant burden on analysts who may face long hours, complex alert queues, and the cognitive pressure of distinguishing real threats from benign variations under time-sensitive conditions.

The human-on-the-loop model shifts the balance toward greater autonomy for AI systems while retaining human supervision. Here, AI is empowered to take certain predefined actions automatically, particularly for threats that are routine, well-understood, or assessed as low risk based on consistent and predictable indicators. This approach dramatically increases response speed and allows analysts to focus their expertise on novel or complex threats that require human insight. It is well-suited for activities such as blocking known malicious IP addresses, quarantining files that match established malware signatures, terminating clearly malicious processes, or enforcing protective steps in response to high-confidence indicators. However, the human role becomes supervisory rather than operational. Analysts must understand the AI's authority, remain prepared to intervene when actions deviate from expected patterns, and conduct periodic reviews to ensure that automated decisions remain aligned with organizational priorities. While effective, this model carries the inherent risk that an incorrectly calibrated automated action may disrupt legitimate activity, affecting customers, partners, or internal stakeholders. Therefore, human-on-the-loop oversight requires trained personnel who are vigilant, empowered, and fully aware of when and how to intervene.

Most modern organizations find that neither model is sufficient on its own. Instead, hybrid approaches provide the flexibility needed to address varying levels of risk and ambiguity. In hybrid systems, AI may act autonomously for low-risk events while escalating uncertain or high-impact events for human evaluation. This dynamic allocation of responsibility enables organizations to move swiftly where appropriate while preserving human control where necessary. Such an adaptive model must be guided by risk appetite, operational sensitivity, and a thoughtful understanding of where automation can accelerate protection without creating undue risk. It also requires clear escalation paths, decision thresholds, and communication mechanisms to ensure that transitions between automated and human decision-making are seamless and predictable.

Central to all of these models is the question of accountability. As AI becomes more woven into the fabric of cybersecurity operations, determining responsibility for decisions becomes increasingly complex. When an automated action blocks legitimate business communication, halts a critical workflow, or fails to identify an attack, determining who bears responsibility cannot be left to chance. AI systems cannot be held accountable for their actions. Responsibility instead lies with the humans and teams who design, deploy, monitor, and supervise AI-driven workflows. This requires organizations to establish clear governance frameworks that define decision rights, delineate responsibilities, and ensure that every automated action is traceable to an accountable human owner.

Governance frameworks must begin with clearly articulated policies that specify the intended use of AI, the degree of autonomy granted to AI systems, and the boundaries that AI must operate within. Policies should clarify what types of decisions AI can make, when it must defer to human judgment, and how overrides should occur. Equally important is the establishment of unambiguous roles and responsibilities across the AI lifecycle, from training and validation to deployment and continuous monitoring. Every action an AI system takes should be backed by a transparent, auditable trail that shows not only what occurred but why it occurred and who had oversight responsibility at each stage.

Audit logging is vital in this regard. Comprehensive system logs provide transparency into automated actions, human approvals, overrides, and the contextual factors behind decision outcomes. These logs enable forensic analysis, compliance reporting, and organizational learning, all of which contribute to continuous improvement and risk mitigation. Regular performance reviews are also essential. These reviews assess whether AI systems are performing as expected, whether decision thresholds are appropriately calibrated, and whether oversight mechanisms remain effective. As both AI systems and threat landscapes evolve, periodic recalibration ensures that automation remains aligned with operational reality and organizational priorities.

Training and competency development further reinforce accountability. Analysts and supervisors interacting with AI systems must be equipped with the necessary skills to identify anomalies, intervene when needed, and apply judgment in complex or ambiguous situations. Training must emphasize not only technical proficiency but also ethical reasoning, risk evaluation, and scenario-based decision-making. Well-developed human expertise ensures that oversight remains strong, even as AI systems take on increasing operational responsibility.

The principle of meaningful human control underpins all of these efforts. This principle asserts that humans must retain the authority to guide and govern AI systems, intervene decisively when necessary, and remain accountable for the outcomes of AI-driven actions. Meaningful human control is essential not only for operational safety but also for ethical integrity. It ensures that AI remains a tool in the service of human judgment rather than an independent operator shaping outcomes without accountability. It safeguards against the possibility that automated systems could inadvertently introduce bias, cause operational harm, or take actions inconsistent with organizational values.

For CISOs and boards, these concepts carry significant leadership implications. Human oversight must be recognized as a fundamental requirement of responsible AI deployment, not an optional safeguard. Decision rights and accountability structures must be clearly defined, documented, and regularly reviewed to ensure that responsibility remains human and traceable. Automation strategies must reflect organizational risk appetite, ensuring that AI autonomy is granted only where appropriate and monitored closely. Governance frameworks must evolve alongside AI capabilities, incorporating periodic evaluations and recalibrations. Workforce development must be prioritized to ensure that personnel remain capable of supervising, interpreting, and intervening in AI workflows. Above all, AI systems deployed without responsible human oversight pose strategic and operational risks, while systems governed by strong leadership, accountability, and disciplined processes enhance resilience, accelerate decision-making, and reinforce organizational trust.

Ultimately, the future of cybersecurity will be shaped not by the extent to which organizations automate but by the wisdom with which they balance automation with human oversight. AI brings speed, scale, and analytical depth, yet it is human judgment, ethical consideration, and accountability that secure the enterprise. Organizations that strike this balance will leverage AI as a powerful ally. Those that fail to do so may find themselves governed by systems they cannot fully control or understand. Responsible leadership ensures that AI strengthens security, preserves trust, and supports a resilient, ethically grounded cybersecurity posture.

Developing an Ethical AI Charter for Security Teams

As artificial intelligence becomes woven into the fabric of modern cybersecurity operations, its adoption must be accompanied by a well-defined ethical foundation. The pace of innovation in AI-driven security tools has outstripped the development of standardized ethical frameworks, leaving many organizations unsure of how to govern technologies that can make decisions at machine speed with real-world implications for users, customers, partners, and the business itself. For CISOs, security executives, and boards, this gap represents more than a procedural oversight; it is a strategic risk. Without explicit ethical guidelines, AI-enabled security programs risk drifting into unintended surveillance, discriminatory outcomes, overreliance on opaque systems, and inconsistent decision-making that undermines trust. To prevent such pitfalls, security organizations must formalize their principles and expectations through a carefully constructed Ethical AI Charter that governs the design, deployment, and operation of AI technologies within the security function.

An Ethical AI Charter is not another policy for compliance shelves; it is a strategic declaration of how an organization will balance innovation with responsibility. It reflects the values that shape security culture and establishes the moral boundaries within which AI must operate. It provides clarity not only to technical teams but also to executives, regulators, employees, and external stakeholders who expect assurances regarding fairness, accountability, privacy, and the ethical stewardship of data. As a living document, the charter evolves alongside advances in AI capabilities and the threat landscape. It must be reviewed and refined regularly, ensuring that ethical governance keeps pace with technological transformation rather than lagging behind it. The commitment to revisit and reassess the charter is a signal of organizational maturity, demonstrating that ethics is not an afterthought but an ongoing discipline integral to security strategy.

The development of an Ethical AI Charter begins with reaffirming the organization's mission and values. Security teams must articulate how AI should contribute to the protection of people, systems, and data, and how ethical obligations complement and strengthen that mission. An organization that places high value on privacy, for example, must ensure that AI systems do not erode that principle through excessive monitoring or unnecessary data collection. These foundational values become the lens through which all AI deployment decisions are evaluated. They must be explicit so that both technical and nontechnical stakeholders can understand the philosophical commitments driving the organization's approach to AI. Too often, organizations deploy AI tools reactively to address immediate tactical challenges without considering broader ethical implications. A charter brings intention to the process, requiring decision-makers to evaluate not only what AI can do but also what it should do.

A core function of the charter is to establish ethical boundaries and red lines. These boundaries identify where AI's use may expose the organization to heightened ethical risk and define the circumstances under which AI deployment requires stricter oversight or outright prohibition. For example, the charter may affirm that AI cannot be used to infer sensitive characteristics about employees or customers, or to engage in intrusive surveillance beyond legally permissible and operationally justified activities. It may prohibit reliance on AI outputs in decisions that materially affect individuals without human validation, especially in areas such as employee investigations, insider threat detection, or performance assessment. These boundaries protect the organization from unintended harm and ensure that the introduction of AI does not erode organizational values or violate societal norms. By defining red lines in advance, leaders prevent arbitrary or inconsistent decision-making and reinforce that ethical discipline is as important as technical rigor.

The charter also provides a systematic process for navigating ethical dilemmas. Cybersecurity rarely presents decisions that are purely technical or purely ethical. Instead, leaders must often weigh competing priorities such as security and privacy, speed and caution, or autonomy and oversight. AI systems amplify these tensions because they operate at scale and can propagate decisions rapidly. The charter offers a structured method for evaluating such trade-offs, encouraging leaders to consider potential consequences, seek cross-functional input, assess legal implications, and elevate decisions that require broader consultation. By outlining how ethical concerns should be escalated and adjudicated, the charter ensures that decision-making is neither ad hoc nor left solely to technical teams. Instead, it promotes a collaborative model where legal, compliance, HR, security, and executive leadership share responsibility for ethically charged decisions.

Ethical AI governance also hinges on the integrity and stewardship of data. AI is only as ethical as the data that fuels it. A robust charter requires that all data used for training and operational decision-making be collected lawfully, protected rigorously, and reviewed regularly for bias or representational gaps that may lead to skewed or discriminatory outcomes. Where possible, data should be anonymized or minimized to reduce unnecessary exposure of sensitive information. If an AI system is trained to detect insider risk, for example, the charter would require that the data reflect diverse behavioral patterns across departments, job roles, and demographics to avoid systemic bias. Data provenance becomes a critical consideration, ensuring that the organization understands the origins, quality, and limitations of its training datasets before deploying AI tools that may influence real-world decisions.

Transparency must also be embedded into the charter. While not every stakeholder needs deep technical detail, the organization must commit to ensuring that AI systems can be understood and explained at a level appropriate for oversight. Decisions that significantly impact individuals or operations must be justifiable. The charter therefore establishes expectations for documentation, traceability, decision logs, and clarity in system behavior. These commitments build trust within and beyond the organization, enabling regulators, auditors, and stakeholders to evaluate the fairness and appropriateness of AI-driven actions and ensuring that the internal security team can supervise these systems effectively.

Accountability is another defining pillar. The charter must specify who is responsible for the ethical integrity of AI systems across the entire lifecycle. If an AI-enabled tool misclassifies an event and triggers an inappropriate response, the organization needs clarity on which teams or roles are accountable for reviewing, correcting, and learning from the incident. This avoids the dangerous tendency to treat AI outputs as infallible or to allow responsibility to diffuse across teams until no one is clearly accountable. Defined accountability structures ensure that AI deployments remain aligned with policy, that issues are resolved promptly, and that the organization consistently learns from mistakes.

A mature Ethical AI Charter also emphasizes continuous learning and adaptation. As AI systems evolve and new ethical concerns emerge, security teams must remain informed and prepared to adapt. This involves training security professionals on AI capabilities, limitations, ethics, and governance expectations. It encourages a culture where ethical considerations are not treated as compliance hurdles but as core components of operational excellence. Leaders must promote open dialogue on ethical concerns and ensure that personnel feel empowered to question AI-driven processes when something appears inconsistent, biased, or potentially harmful.

The creation of the charter should itself be an inclusive exercise. Ethical decisions that impact security operations cannot be made in isolation. Legal teams bring perspectives on regulatory compliance, privacy experts bring insight into acceptable boundaries, HR helps interpret workforce implications, and ethicists provide depth on societal and organizational values. Involving diverse voices ensures that the charter reflects a comprehensive view of ethical risk rather than the lens of any single discipline. It also ensures broader buy-in, increasing the likelihood that ethical principles are internalized across teams rather than seen as directives imposed from leadership.

Ultimately, the Ethical AI Charter becomes a proactive instrument of risk management, governance, and organizational integrity. It reduces uncertainty, prevents unintended consequences, and provides a structured pathway for ethical decision-making in a domain that is increasingly mediated by intelligent systems. By articulating clear boundaries, defining roles and responsibilities, strengthening data stewardship, and embedding transparency and accountability, the charter not only protects the organization but reinforces its commitment to responsible innovation. Organizations that develop and maintain such charters send a powerful message to regulators, customers, employees, and partners that AI is being adopted with intention, discipline, and respect for human and organizational values.

In a landscape where cyber threats evolve rapidly and AI systems become central to defensive strategy; the Ethical AI Charter stands as the moral anchor of the security team. It ensures that technological power is matched by ethical responsibility and that the pursuit of security never compromises the organization's duty to fairness, privacy, and the trust of those it serves. Approached thoughtfully, the charter strengthens both security outcomes and organizational integrity, securing not only systems but the human values that define the enterprise itself.

Executive Vignette: The Charter That Prevented a Crisis

When Meridian Financial Services began integrating an AI-driven insider risk monitoring system into its security operations, the leadership team viewed it as a powerful leap forward. The system promised the ability to detect anomalous employee behavior far more quickly than traditional controls, identifying early indicators of fraud, data exfiltration, or privilege misuse long before human analysts could. The board approved the deployment, compliance expressed cautious optimism, and the security team prepared for a new era of proactive defense.

Within weeks of deployment, the system generated a high-severity alert involving an employee in the regional lending division. The AI flagged several behaviors as suspicious, including large data downloads, irregular login times, and access to a customer dataset that the employee did not typically use. The automated recommendation suggested immediate credential suspension and initiation of a formal investigation. On the surface, the indicators seemed compelling. Yet the Ethical AI Charter, which Meridian had finalized only months earlier, required that any AI-driven alert with potential disciplinary consequences undergo human review before any action could be taken. The charter also required that all AI-flagged concerns involving personnel be contextualized through business validation, privacy review, and an analysis of operational impact before escalation.

Following the charter's guidelines, the case was routed to a designated review group composed of a senior security analyst, a representative from HR, and a privacy officer. The analyst reviewed the system logs and cross-referenced the behavior with recent project activity. The HR representative confirmed that the employee had recently been assigned to a cross-functional data migration initiative involving the very datasets the AI had flagged. The privacy officer noted that the employee had been granted temporary permissions authorized by their manager, though the change had not yet synchronized across all monitoring systems. The unusual login times aligned with the employee supporting a multi-region project team spread across time zones.

Rather than representing the early stages of an insider threat, the activity was the predictable outcome of a legitimate business project operating on an accelerated timeline. If

Meridian had acted on the AI system's automated recommendation without consulting the Ethical AI Charter, the consequences would have been immediate and severe. The employee would have been unjustly suspended, the project's schedule disrupted, customer data migration delayed, and internal trust eroded. The incident could have escalated into a legal claim for wrongful action, damaging the company's reputation and weakening confidence in both AI systems and security leadership.

Instead, the review concluded that the alert reflected a gap in the AI system's access-change synchronization and a lack of visibility into project-based permissions. The team updated the AI workflow to ingest approved temporary access lists and ensured that change management procedures aligned with monitoring logic. The employee was informed of the review in a transparent and respectful manner, reinforcing the organization's commitment to fairness rather than secrecy. The incident became a case study in how governance, if practiced consistently, prevents harm rather than merely responding to it.

For Meridian's leadership, the outcome validated the purpose of the Ethical AI Charter. It demonstrated that the charter was more than a document; it was a mechanism that safeguarded employees, protected the organization from avoidable risk, fostered trust in AI-enabled systems, and ensured that security decisions reflected both efficiency and integrity. The board later cited the incident as evidence that disciplined ethical oversight strengthens cybersecurity outcomes not by slowing innovation but by grounding it in principled, responsible practice. Leadership recognized that the organization had avoided a reputational and operational crisis not because the AI system was perfect, but because governance ensured that humans remained responsible stewards of powerful technology.

Conclusion: Ethics as a Security Control, Responsibility as a Leadership Standard

AI-driven security only earns the right to scale when it earns the right to be trusted. This chapter has reinforced that responsible AI deployment is not a philosophical add-on to technical effectiveness. It is a core control surface that determines whether AI strengthens enterprise defense or silently introduces new forms of harm, inequity, and governance failure. As AI systems increasingly detect threats, influence prioritization, and initiate action, cybersecurity leadership must treat ethical responsibility as inseparable from operational resilience.

The ethical imperative becomes concrete through the risks explored here: bias embedded in data and decision logic, opacity that undermines explainability and auditability, accountability gaps created by automation, and privacy erosion through overcollection or intrusive monitoring. The Meridian vignette illustrates the real-world cost of neglecting these controls. When systems act without transparency, oversight, or fairness safeguards, damage propagates beyond the SOC into business continuity, legal exposure, employee trust, and patient or customer impact. This is why ethical performance must be managed with the same rigor applied to technical performance.

Responsible AI deployment requires intentional structure. Bias must be identified and mitigated through disciplined data stewardship, subgroup testing, continuous monitoring, and operational feedback loops. Explainability must be built into acquisition and implementation standards so security teams can validate decisions, reduce harm, and support regulatory expectations. Human oversight must remain a deliberate design choice, not an afterthought, with clear decision thresholds, override authority, and auditable accountability across the AI lifecycle. The Ethical AI Charter operationalizes these expectations by turning values into governance, ensuring that AI adoption remains aligned with fairness, transparency, privacy, and organizational integrity.

The leadership mandate is clear. Security executives must govern AI not only for what it can detect, but for how it behaves under pressure, how it impacts people, and how it stands up to scrutiny. Organizations that embed ethics into AI security practices will accelerate adoption with confidence, preserve stakeholder trust, and sustain resilience in environments where automation is unavoidable. Organizations that do not find that AI amplifies risk as efficiently as it amplifies capability. In the AI era, ethical discipline was not a constraint on security leadership. Ethical discipline is what makes advanced security defensible.

Chapter 4: Strategic AI Risk Management for Security

Artificial Intelligence is rapidly transforming cybersecurity, offering unprecedented speed, scale, and analytical power, yet introducing risks that fundamentally differ from traditional technologies. These systems learn from data, evolve over time, and can fail, drift, or be manipulated in ways that are subtle and difficult to detect, creating new vulnerabilities across the organization. As AI becomes deeply embedded in security operations, leaders must recognize that managing AI risk is not a technical exercise but a strategic imperative that spans governance, compliance, operations, and resilience. This chapter provides a comprehensive framework for understanding AI's unique risk landscape, integrating those risks into enterprise governance structures, continuously monitoring AI performance, defining measurable Key Risk Indicators, and developing incident response plans tailored to AI-specific failure modes. The goal is to equip security executives and boards with the clarity, structure, and oversight mechanisms needed to deploy AI responsibly and confidently within mission-critical environments.

The Unique Risk Landscape of AI in Cybersecurity

The integration of Artificial Intelligence (AI) into cybersecurity introduces unprecedented capabilities while simultaneously creating risks that extend beyond traditional IT systems. AI models represent dynamic, data-driven constructs whose behaviors evolve over time, making them vulnerable in ways that static software is not. These systems bring new attack surfaces, ranging from model manipulation to subtle undermining of their learning processes, each capable of turning a defensive asset into a liability if compromised. The same intelligence that enhances detection, prediction, and automation can be weaponized against the organization when integrity is lost. Because of this dual potential, security leaders must understand and manage AI-specific risks as a strategic requirement, not an optional enhancement. This requires a shift in mindset, recognizing that AI systems operate within a dynamic environment where threats evolve and vulnerabilities may emerge from any stage of the AI lifecycle.

One of the most critical risks stems from the AI model itself, whose integrity can be compromised during development, training, or deployment. Unlike traditional applications defined by deterministic code, AI models rely on statistical patterns learned from data, making them susceptible to subtle manipulations. A compromised model may misclassify benign traffic as malicious, raising operational friction, or fail to detect genuine threats altogether, creating exploitable blind spots. Safeguarding model integrity therefore requires rigorous validation, restricted access to training environments, secure model repositories, and protections that account for how AI decision boundaries can shift under tampering. By treating the model as a high-value asset throughout its lifecycle, organizations can better anticipate and detect attempts to alter its behavior.

Data poisoning represents another powerful attack vector because corrupting training data can distort the AI's understanding of what constitutes safe or malicious behavior. Attackers may inject mislabeled or adversarial samples into training sets to bias the model to overlook specific threat patterns. Such degradation can unfold gradually, making it difficult to attribute performance issues to poisoned data rather than natural drift. In practice, data poisoning may cause a phishing-detection model to label malicious messages as safe or a fraud-detection model to tolerate suspicious transactions. Effective defense requires strong data provenance, anomaly detection within training pipelines, and periodic validation against trusted reference datasets to reveal inconsistencies in learned patterns.

Adversarial attacks pose an operational threat by subtly manipulating input data to produce incorrect AI predictions while remaining invisible to human observers. This is particularly dangerous because attackers can craft malicious payloads that evade deep learning models without altering core functionality. Minor perturbations to malware binaries, network packet structures, or visual features can cause misclassification while appearing benign. These examples illustrate a fundamental characteristic of many machine learning models—they may be brittle when confronted with inputs outside their learned distributions. Defense strategies such as adversarial training, ensemble models, and input sanitization can increase robustness, but the risk remains inherent until intentionally mitigated.

Concept drift presents yet another challenge as the threat landscape evolves and user behaviors shift over time. AI models trained on historical data may become less effective as legitimate processes change or attackers introduce new techniques. Over time, the model's performance degrades as reality diverges from its original assumptions, leading to higher false positives or false negatives. Detecting drift requires continuous monitoring of key performance indicators and retraining models to realign them with emerging patterns. Without an intentional MLOps strategy for detecting and addressing drift, organizations risk deploying increasingly outdated and ineffective AI defenses.

AI deployment also introduces risks related to data sensitivity, system complexity, and supply chain dependencies. These systems typically process large volumes of sensitive information, making them attractive targets for data exfiltration or manipulation. The interconnected pipelines required for data ingestion, model training, and inference create new dependencies that can be exploited by attackers or affected by upstream failures. Third-party AI components may also introduce supply chain vulnerabilities that propagate into operational environments. Robust access controls, comprehensive supply chain vetting, and ongoing vulnerability assessments are essential to mitigating these risks and preventing cascading failures.

AI's complexity can also lead to emergent or unintended behaviors that differ from what developers anticipate. These behaviors may arise from unmodeled interactions between the model and its environment, resulting in excessive false positives, unexpected actions, or operational disruptions. In some cases, poorly designed AI systems may even exhibit self-optimizing or competitive behaviors that stray from intended outcomes. Testing AI systems in diverse and simulated environments helps uncover these potential failures before deployment. Human oversight remains essential to identifying and mitigating behaviors that are misaligned with operational expectations.

Explainability poses another significant challenge because many advanced AI models operate as "black boxes" whose internal logic is difficult to interpret. When these systems drive security actions; such as account lockouts, transaction blocks, or automated containment, the inability to fully audit decisions can create legal, operational, and compliance issues. A lack of transparency also complicates incident response and forensic analysis because analysts may struggle to determine why a model behaved incorrectly. Pursuing explainable AI (XAI) techniques and maintaining documentation of model rationale supports both compliance and operational trust, especially in regulated environments.

Finally, an over-reliance on AI-based automation can create systemic vulnerabilities if human oversight diminishes or skills atrophy. While AI can streamline workflows and enhance detection, it cannot fully replace human judgment in high-stakes and novel scenarios. Excessive dependence on AI may lead analysts to overlook anomalies or fail to intervene effectively during system failures. Maintaining an appropriate balance between automation and human expertise ensures resilience and prevents the organization from becoming vulnerable when AI encounters unfamiliar threats. Training programs, collaborative decision workflows, and hybrid human-AI review processes are essential components of a sustainable approach.

Integrating AI Risks into the Enterprise Risk Management ERM Framework

Integrating AI risks into the broader Enterprise Risk Management (ERM) framework ensures that AI-specific vulnerabilities are not treated as isolated technical issues but as strategic organizational risks. Traditional ERM models focus on operational, financial, compliance, and cybersecurity risks, yet AI introduces cross-cutting impacts that span all these categories. Proper integration provides executives with a unified view of AI-related exposures and aligns AI security practices with established governance processes. This approach ensures that risk appetite, resource allocation, and oversight mechanisms account for AI's unique characteristics and potential systemic effects. Treating AI risks holistically strengthens the organization's ability to identify cascading impacts and prioritize mitigation in accordance with strategic objectives.

The integration process begins with a comprehensive mapping of AI risks across the entire lifecycle from data acquisition and preparation to model development, deployment, and decommissioning. Each stage contains distinct risks, including data bias, privacy violations, infrastructure vulnerabilities, adversarial attacks, and model drift. Mapping these risks clarifies how AI system failures can propagate into broader operational, compliance, or financial consequences. For example, biased models may trigger regulatory scrutiny, while adversarial attacks may cause widespread operational disruption. By translating AI-specific technical concerns into enterprise-level risk categories, organizations ensure that AI risks are visible to risk committees, audit teams, and executive leadership.

Assessing the likelihood and impact of AI risks requires adapting existing ERM methodologies to accommodate the uncertainty inherent in AI systems. Because historical data may not accurately predict the likelihood of emerging AI-specific attacks, assessments must incorporate scenario modeling, expert judgment, and threat intelligence. Impact assessments must also extend beyond direct financial effects to include reputational harm, customer trust erosion, and ethical consequences. Using standardized risk matrices while allowing for AI-specific variables ensures consistency with ERM practices while capturing novel dimensions of AI vulnerability. A multidisciplinary assessment process engaging data scientists, security leaders, legal teams, and business stakeholders ensures a comprehensive risk perspective.

Developing risk treatments within the ERM framework involves applying traditional risk approaches; avoidance, mitigation, transfer, and acceptance through the lens of AI's particular challenges. Risk avoidance may be appropriate when AI applications introduce unacceptable ethical or operational risks that cannot be reasonably mitigated. For most AI deployments, mitigation strategies form the core of risk treatment and include technical controls such as adversarial defenses, rigorous data validation, model auditing, and secure MLOps processes. Governance controls, including ethics reviews, transparency guidelines, and AI development policies, further support responsible deployment. In some cases, risk transfer through insurance products may be possible, though organizations must carefully evaluate policy limitations related to AI failures or algorithmic bias. Risk acceptance requires documented justification, alignment with risk appetite, and ongoing monitoring to ensure conditions do not change.

Ongoing monitoring and review of AI risks are essential because AI systems operate in dynamic environments where threats, data patterns, and operational contexts evolve. The ERM framework must incorporate mechanisms to continuously evaluate the effectiveness of AI controls and the relevance of risk assessments. Key Risk Indicators (KRIs), discussed later in this chapter, provide quantifiable measures that support ongoing evaluation. Regular audits, stress tests, and performance reviews ensure that AI risks remain visible and appropriately managed. Embedding AI risk metrics into quarterly risk committee reporting ensures that leadership maintains awareness of emerging trends and vulnerabilities.

Integrating AI risks into ERM also requires cultivating a risk-aware culture throughout the organization. AI risks impact not only cybersecurity but product development, customer service, compliance, and strategic decision-making. Training, communication, and clear role definitions help ensure that AI risk responsibilities are well understood and consistently applied. Updating ERM documentation, including risk registers, risk appetite statements, and internal controls to explicitly address AI ensures alignment with governance standards. This integration ensures that AI risk management becomes a routine component of audits, board reviews, and compliance reporting.

Strong governance and oversight are essential to ensuring that AI risk management remains effective. Boards and senior leadership must understand AI's strategic implications and ensure that appropriate resources and structure exist to manage its risks. Accountability for AI risk should be clearly assigned, with defined escalation paths and periodic reporting requirements. When properly integrated, AI risk management within the ERM framework enables organizations to embrace

AI innovation while maintaining responsibility, transparency, and resilience.

Continuous Monitoring of AI Models and Performance

Once deployed, AI systems must be continuously monitored because they operate in dynamic environments where data patterns, user behaviors, and threat actors constantly evolve. Unlike traditional security tools that operate predictably, AI models can degrade, drift, or be manipulated over time without obvious signs. Continuous monitoring acts as an early warning mechanism, identifying performance degradation, adversarial interference, operational bottlenecks, or shifts in input data distributions. Without such monitoring, AI systems may silently fail, creating blind spots that undermine the organization's security posture. Effective monitoring ensures that AI models remain aligned with intended outcomes and adapt to the changing threat landscape.

Model drift is a primary driver for continuous monitoring, occurring when the statistical properties of the model's inputs or target outputs shift over time. Concept drift changes the meaning of what the model must detect, while data drift alters the distribution of features that inform predictions. For security systems, attackers may adopt new techniques, employees may adopt new workflows, or business processes may evolve, leading to gradual model performance decay. Drift may manifest as rising false positives, declining recall, or increased uncertainty in predictions. Monitoring these indicators allows organizations to detect issues early and trigger retraining or recalibration before performance degradation affects critical operations.

Adversarial manipulation represents another critical monitoring priority because AI models can be intentionally deceived by crafted inputs designed to cause misclassification. Unlike traditional exploits, adversarial attacks may not involve system compromise; instead, they exploit model vulnerabilities directly. These attacks may appear as subtle perturbations in input data that produce disproportionate effects on model outcomes, often eluding conventional detection. Monitoring anomalies input patterns, latent representations, and model sensitivity helps reveal adversarial behavior. By identifying unusual prediction instability or sudden shifts in feature importance, organizations can detect attempts to manipulate AI systems in real time.

Effective continuous monitoring requires dashboards that present real-time metrics reflecting the health and reliability of AI models. Performance metrics such as accuracy, precision, recall, and F1-score provide essential insights, but trending over time yields the most diagnostic value. Confidence score distributions reveal model certainty levels and can expose drift or emerging uncertainty. Monitoring the statistical properties of incoming data ensures that deviations from the training distribution are detected early. Feature importance monitoring helps identify unexpected behavioral changes in explainable models. Integrating fairness and bias metrics ensures that the AI's decisions remain equitable and compliant with regulatory expectations, reducing ethical and legal risks.

Operational metrics also play a crucial role in continuous monitoring. Model latency and throughput provide insights into the system's ability to process data efficiently and may reveal resource contention or performance bottlenecks. Reconstruction errors in models like autoencoders may indicate anomalies or adversarial inputs. Monitoring computational resource utilization helps identify inefficiencies, denial-of-service attempts targeting AI infrastructure, or unanticipated resource drains. Aggregating these metrics into role-specific dashboards ensures that data scientists, SOC analysts, and business leaders each receive relevant insights tailored to their responsibilities.

Essential performance indicators include:

Accuracy/Precision/Recall/F1-Score: These standard machine learning metrics remain critical, but their trending over time is what provides true value. A gradual decline in accuracy or a significant drop in recall (the model's ability to find all relevant instances) can be early warning signs of drift. For a fraud detection system, a declining recall means more fraudulent transactions are slipping through.

Confidence Scores: Most AI models provide a confidence score with their predictions. Monitoring the distribution of these scores is crucial. A decrease in the average confidence score across all predictions, or an increase in low-confidence predictions, can indicate that the model is becoming less certain in its outputs, a sign that it may be encountering data outside its training distribution or experiencing drift.

Data Distribution Analysis: Tools should be in place to continuously compare the statistical distribution of live input data against the distribution of the training data. Deviations in mean, variance, or the presence of new categories in the input can signal data drift. Techniques like Kullback-Leibler divergence or Jensen-Shannon divergence can quantify these differences.

Feature Importance Monitoring: For explainable AI models, tracking the relative importance of different input features can reveal shifts. If a feature that was previously a strong predictor suddenly diminishes in importance, or a new, unexpected feature becomes dominant, it suggests that the underlying relationships the model learned are changing.

Bias and Fairness Metrics: It is imperative to continuously assess whether the AI model exhibits any unintended bias across different demographic groups or other sensitive attributes. Monitoring metrics like demographic parity, equalized odds, or predictive equality helps ensure that the AI's decisions remain fair and compliant with ethical and regulatory standards. Any widening gap in these metrics indicates a need for intervention.

Adversarial Attack Detection Metrics: This is a more challenging area, but monitoring for anomalies that are characteristic of adversarial attacks is vital. This can include:

Input Perturbation Analysis: Detecting unusual levels of minor modifications to input data that precede a misclassification.

Outlier Detection on Latent Representations: Analyzing the internal representations the AI creates for data points. Adversarial examples often map to unusual regions in this latent space.

Prediction Stability: Monitoring how sensitive predictions are to minor changes in input. Highly sensitive predictions can be a red flag.

Reconstruction Error: For models that involve reconstruction (e.g., autoencoders used for anomaly detection), monitoring the error in reconstructing input data. Unusually high reconstruction errors might indicate manipulated inputs.

Model Latency and Throughput: While not directly related to accuracy, significant increases in processing time or decreases in the number of requests handled per unit of time can indicate underlying issues, potentially including computationally intensive adversarial attempts or an overloaded system struggling with new data patterns.

Alerting mechanisms support timely response by identifying deviations that exceed established thresholds. Alerts must be dynamic and context-aware, distinguishing between normal fluctuations and meaningful anomalies that require investigation. High-priority alerts might include sudden increases in false negatives, evidence of adversarial perturbations, or large shifts in data distributions affecting model reliability. Lower-priority alerts may indicate gradual drift or minor data pipeline issues that can be resolved through routine maintenance. Integrating AI-related alerts into broader SIEM or SOAR systems enables correlation with other security events and supports coordinated incident response.

Continuous monitoring should be embedded within MLOps pipelines to ensure that monitoring, maintenance, and model improvement form a cohesive operational workflow. Automated data validation prevents corrupted or malformed data from reaching the model, reducing the risk of poisoning or unexpected drift. Automated retraining pipelines can be triggered when performance thresholds fall below acceptable levels, ensuring that models remain aligned with current conditions. Canary deployments and A/B testing support safe rollout of updated models and ensure that performance improvements are validated before full deployment. These processes maintain model accuracy while safeguarding against unintended consequences.

This includes:

Automated Data Validation Pipelines: Ensuring that incoming data is validated against predefined schemas and statistical properties before it even reaches the AI model. Any anomalies are flagged immediately.

Automated Model Retraining Triggers: When monitoring systems detect significant drift or performance degradation, they should be capable of automatically triggering a retraining process for the AI model, using updated data and potentially revised model architectures. This closes the loop on drift and ensures the model remains relevant.

Canary Deployments and A/B Testing for Model Updates: When a retrained or new model is ready, it should not be deployed to the entire user base immediately. Canary deployments (releasing to a small subset of users) or A/B testing (running the new model alongside the old one for a period) allow for real-world validation of the updated model's performance and stability before a full rollout. Monitoring plays a crucial role in evaluating the results of these tests.

Feedback Loops from Human Review: For systems that incorporate a human-in-the-loop component, the feedback provided by human reviewers is invaluable. This feedback should be systematically collected, analyzed, and used to both identify model errors and to refine the monitoring and alerting systems themselves. If human reviewers consistently flag certain types of misclassifications, it indicates an area where the model is failing and where monitoring might need to be enhanced.

Human feedback loops remain critical in continuous monitoring because human analysts can provide high-quality labels and contextual insights that further refine model performance. Feedback from analysts in human-in-the-loop systems can identify blind spots, reveal emerging threat patterns, and support model retraining efforts. Integrating this feedback into monitoring dashboards ensures that human insights influence ongoing model evaluation. By combining human expertise with automated monitoring, organizations strengthen resilience and maintain higher trust in AI-driven security systems.

Establishing Measurable Key Risk Indicators KRIs for AI

Key Risk Indicators (KRIs) translate raw monitoring data into actionable insights that support proactive AI risk management. While traditional cybersecurity metrics remain relevant, AI demands more specialized indicators that capture the nuanced behaviors, vulnerabilities, and ethical considerations of machine learning systems. KRIs help organizations understand whether AI systems are performing within acceptable risk thresholds and provide early warnings before failures escalate. By aligning KRIs with the organization's risk appetite and security objectives, CISOs ensure that AI-driven systems are continuously evaluated for reliability, fairness, and resilience.

Model performance and accuracy degradation form a foundational category of AI-specific KRIs. Tracking trends in accuracy, precision, recall, and F1-score over time reveals whether the model is drifting or losing effectiveness. KRIs should define acceptable performance thresholds and outline escalation levels when metrics fall below expectations. False positive and false negative rates are particularly important for security applications because deviations may indicate operational inefficiencies or vulnerabilities. Monitoring changes in confidence scores and their distribution helps identify increased uncertainty or shifts in data quality. These KRIs enable early intervention before degraded performance affects security outcomes.

Effectiveness against evolving threats represents another essential category of KRIs. AI systems often detect anomalies or novel attack vectors that traditional tools miss, making their adaptability crucial. KRIs may include the model's success rate in identifying new threat types, performance in controlled red team exercises, or resilience to adversarial perturbations. Time-to-detect and time-to-respond metrics for AI-flagged incidents reflect the system's operational impact and its contribution to the incident response process. When performance in these areas declines, KRIs signal a need to retrain the model, refine detection pipelines, or strengthen adversarial defenses.

Ethical and compliance integrity must also be monitored through KRIs because AI systems can inadvertently produce biased or discriminatory outcomes. KRIs may track disparities in error rates across demographic groups, privilege levels, or customer segments. Monitoring these indicators helps organizations identify ethical risks and comply with regulatory requirements related to fairness and transparency. Additional KRIs evaluate adherence to privacy regulations, such as the effectiveness of data anonymization or compliance with retention policies. These indicators ensure that AI systems operate responsibly and maintain trust among users, regulators, and stakeholders.

Operational efficiency and resource utilization are also critical categories of KRIs for AI deployments. AI models require substantial computational resources, and inefficiencies or anomalies may reveal deeper system issues. KRIs may track GPU utilization, memory consumption, or model inference latency to identify bottlenecks or potential denial-of-service attempts targeting AI infrastructure. Monitoring data pipeline health ensures that data ingestion remains reliable and compliant with expected performance metrics. Retraining frequency KRIs help determine whether models are being updated too often or too infrequently, signaling instability or insufficient adaptability.

Implementing KRIs requires a structured approach that includes contextualization, baseline establishment, threshold definition, automated data collection, and periodic review. Each KRI should directly support the organization's security objectives and reflect meaningful indicators of AI risk. Baselines derived from historical performance help define realistic expectations for normal behavior. Actionable thresholds enable automated alerting and escalation workflows. Continuous refinement ensures that KRIs remain aligned with changes in business context, threat landscapes, and model updates. Integrating AI KRIs into existing ERM and governance processes ensures visibility at the executive and board levels.

For example, an AI fraud detection system might define KRIs such as a minimum 30-day rolling recall of 99.5%, allowable false positive rates under 0.1%, required detection of a set number of new fraud typologies, acceptable fairness deviations across customer segments, and transaction scoring latency under 50 milliseconds. Tracking these indicators ensures that the system maintains its intended performance, ethical alignment, and operational reliability. As organizations rely more heavily on AI for critical security functions, KRIs become indispensable tools for safeguarding against drift, degradation, or unintended consequences.

The implementation of these KRIs requires a structured approach:

1. **Contextualization and Prioritization:** Each KRI must be tailored to the specific AI system and its role within the organization's security architecture. What is a critical KRI for an AI-powered SOC analytics platform might be less important for an AI that automates HR onboarding compliance checks. Prioritize KRIs based on their direct impact on the organization's most significant AI-related risks.

2. **Baseline Establishment:** For each KRI, establish a clear baseline performance metric derived from historical data or initial deployment performance. This baseline serves as the reference point for detecting deviations and changes.

3. **Threshold Definition and Alerting:** Define clear, actionable thresholds for each KRI that trigger alerts at different severity levels. These thresholds should be realistic and based on the organization's risk appetite. Alerts should be routed to the appropriate teams for investigation and remediation. For instance, a minor deviation might trigger a low-priority ticket for the ML operations team, while a

significant breach of a critical KRI might escalate to a full security incident response.

4. Data Collection and Automation: Ensure that the necessary data for calculating each KRI is collected automatically and reliably. This often involves leveraging MLOps platforms, logging mechanisms, and dedicated monitoring tools. Manual calculation of KRIs for AI systems is often impractical and prone to error.

5. Regular Review and Refinement: KRIs are not static. The threat landscape evolves, AI models are updated, and the organization's risk posture changes. Therefore, KRIs and their associated thresholds must be reviewed and refined periodically, ideally on a quarterly or semi-annual basis, or in response to significant incidents or system changes. This ensures they remain relevant and effective.

6. Integration with Governance Frameworks: KRIs for AI should be integrated into the broader enterprise risk management and IT governance frameworks. This ensures that AI risk management is not an isolated activity but a core component of the organization's overall risk management strategy. Reporting on AI KRIs should be a standard agenda item in risk committee meetings.

Consider an AI system designed to detect fraudulent financial transactions. Its KRIs might include:

Detection Rate (Recall) Trend: A KRI could be that the 30-day rolling average recall must not drop below 99.5%. A drop to 99.2% triggers a Level 1 alert, requiring an analysis of recent transaction patterns. A drop to 98.5% triggers a Level 2 alert, initiating an urgent investigation and potential model rollback.

False Positive Rate Threshold: The daily average FPR must remain below 0.1%. A sustained daily average exceeding 0.15% for more than 48 hours leads to a Level 1 alert for investigation into potential system noise or misconfigurations.

New Fraud Pattern Detection Rate: The AI must successfully flag at least 5 previously unseen fraud typologies per month, validated by fraud analysts. A shortfall here indicates a potential gap in the AI's adaptability to emerging threats.

Bias in Fraud Scoring: A KRI could monitor that the average fraud score assigned to transactions originating from different geographic regions or customer segments does not deviate by more than 10% from the overall average, ensuring fairness and preventing discriminatory outcomes.

Model Response Latency: The time taken for the AI to score a transaction must remain below 50 milliseconds for 99% of transactions. Significant latency increases could impact transaction processing speed and user experience and potentially indicate resource contention or an attack.

By establishing and diligently monitoring these types of KRIs, CISOs can move beyond simply deploying AI for security to actively managing the risks inherent in its operation. These measurable indicators provide the clarity and foresight necessary to ensure that AI systems not only perform as intended but also contribute positively and securely to the organization's overall resilience in the face of an ever-evolving threat landscape. They are the vital signs of our AI security investments, allowing us to diagnose problems early, confirm the efficacy of our defenses, and maintain confidence in technology's strategic value.

Developing Incident Response Plans for AI-Related Failures

Incident response plans must evolve to address the unique failure modes and vulnerabilities associated with AI systems. Traditional incident response frameworks often assume

112

deterministic system behavior and well-defined attack vectors, but AI introduces probabilistic outputs, dynamic learning processes, and susceptibility to novel forms of manipulation. A comprehensive AI-specific incident response plan anticipates how AI systems might fail, how these failures could manifest, and what specific containment and recovery steps are required. Treating AI as a first-class citizen within the incident response lifecycle ensures resilience, rapid restoration, and consistent alignment with operational priorities.

Identifying potential failure modes is the first step in developing AI-specific incident response procedures. AI model drift may cause performance to deteriorate gradually, resulting in missed detections or excessive false positives. Compromised model integrity through data poisoning or adversarial attacks may cause the model to produce manipulated outputs that benefit attackers. Operational failures such as data pipeline outages, infrastructure overload, or version control errors can disrupt the AI's ability to function reliably. Recognizing these categories helps teams anticipate different types of incidents and tailor response strategies accordingly.

Reliable detection of AI-related failures requires integrating monitoring systems and KRIs into incident response triggers. When thresholds are breached such as a sudden spike in false negatives, major shifts in input distributions, or evidence of adversarial perturbations alerts must escalate according to severity. Incident response plans should define clear escalation paths and assign responsibilities across the SOC, MLOps teams, data scientists, and relevant business units. Coordinated investigations ensure that both operational impact and underlying model behavior are assessed together to produce a comprehensive understanding of the failure.

Containment procedures for AI incidents often require isolating the affected model or disabling automated actions until the root cause is identified. In high-stakes environments, switching the model into a monitoring-only mode prevents it from executing potentially harmful decisions while still capturing data for analysis. In severe cases, fully removing the model from production and reverting to a fallback system whether automated or human-driven may be necessary. The incident response plan must balance maintaining security coverage with preventing the AI's faulty behavior from disrupting operations or enabling attacker advantage.

Eradication and recovery processes focus on addressing the root cause of the failure and restoring the AI system to reliable operation. If the issue results from drift, retraining the model with updated data may restore accuracy. If data poisoning or adversarial interference is identified, compromised datasets must be removed, vulnerabilities must be patched, and the model may require full retraining on verified data. Post-restoration validation through red teaming, canary deployment, or simulation ensures that the model's behavior has stabilized before returning it to full production. Documented acceptance criteria ensure that restored models meet operational requirements and compliance obligations.

Root cause analysis for AI incidents can be exceptionally complex due to the opaque nature of many machine learning models. Incident response procedures should therefore include AI-specific forensic guidelines, such as preserving model states, logging training inputs and outputs, and maintaining versioned records of model parameters. Tools for explainability and interpretability become critical during investigation because they help analysts clarify why the AI made incorrect decisions and identify which factors contributed to the failure. Thorough documentation ensures that the incident contributes to organizational learning and regulatory compliance.

Communication and documentation requirements must be explicitly defined in AI incident response plans. Internal stakeholders need clear updates on the status, severity, and operational impact of AI-related incidents. External communication may be required if failures affect customers, regulators, or partners. Ethical and legal considerations such as bias, privacy violations, or regulatory breaches must be escalated to appropriate oversight bodies, including legal counsel or ethics committees. Comprehensive documentation supports compliance, insurance claims, and continuous improvement.

AI incident response planning should be reinforced through regular tabletop exercises and simulations designed specifically to test AI-related scenarios. These exercises reveal gaps in monitoring, communication, and escalation procedures and help teams become familiar with the complexities of AI failures. By iteratively refining incident response plans based on real-world performance and emerging AI threats, organizations strengthen resilience and preserve trust in AI-driven security operations. Preparing for AI failures with the same rigor applied to traditional cyber incidents ensures that AI enhances security rather than introducing unmanaged vulnerabilities.

Board-Ready Brief

Strategic AI Risk Management for Security

Executive Overview

Artificial Intelligence (AI) is rapidly reshaping cybersecurity operations, but it also introduces novel risks that extend far beyond traditional IT vulnerabilities. AI-driven systems are dynamic, data-dependent, and susceptible to manipulation, drift, bias, and emergent behavior. These characteristics demand a strategic, enterprise-level risk management approach. This brief outlines the key risks associated with AI in security, the imperative to integrate these risks into Enterprise Risk Management (ERM), the necessity of continuous monitoring and measurable Key Risk Indicators (KRIs), and the requirement for AI-specific incident response plans.

For boards and senior executives, the central message is clear: AI enhances capability but expands the organization's risk surface. Effective governance requires treating AI as a strategic asset with systemic risk implications, not as another security tool.

1. The Unique Risk Landscape of AI in Cybersecurity

AI systems differ from traditional security technologies because they learn from data rather than follow deterministic rules. This creates distinct vulnerabilities that require board-level oversight.

Key AI-Specific Risk Categories

- **Model Integrity Risks:**
  AI models can be manipulated or corrupted during development, training, or deployment, leading to incorrect classifications, blind spots, or dangerous automated responses.
  Impact: Operational disruption, security control failure, reputational damage.

- **Data Poisoning Risks:**
  Attackers may introduce malicious or mislabeled data into AI training sets, altering model behavior over time.
  Impact: Reduced detection accuracy, increased false negatives, compromised trust.

- **Adversarial Input Risks:**
  Attackers can subtly alter input data to deceive AI systems without altering functionality.

Impact: Failure to detect malware, evasion of authentication, increased breach risk.

- **Concept & Data Drift:**
  Threat actors change tactics and business processes evolve, causing model performance to degrade.
  Impact: Increased false positives/negatives, operational inefficiencies.

- **Explainability & Accountability Gaps:**
  Black-box AI models may make decisions that cannot be easily audited or justified.
  Impact: Compliance challenges, legal exposure, stakeholder distrust.

- **Privacy Risks & Sensitive Data Exposure:**
  AI systems require large volumes of data; improper handling increases regulatory and reputational risk.
  Impact: Privacy violations, fines, loss of customer trust.

- **Over-Reliance on Automation:**
  Excessive dependence on AI can erode operational expertise and reduce human ability to intervene effectively.
  Impact: Human skill atrophy, slow recovery during AI outages.

Board Implication: AI introduces systemic risk that can amplify or cascade across business processes. Without governance and oversight, AI becomes an unmonitored control replacing human judgment.

2. Integrating AI Risks into the Enterprise Risk Management (ERM) Framework

AI risk must be embedded into ERM so it can be assessed, mitigated, and monitored with the same rigor as financial, operational, and compliance risks.

Core Integration Activities

1. **Lifecycle Risk Mapping:**
   Identify risks across data acquisition, model development, deployment, monitoring, and retirement.
   Outcome: Clear visibility into where failures originate and how they propagate.

2. **Risk Categorization Across ERM Domains:**
   Align AI risks with strategic, operational, compliance, reputational, and financial categories.
   Outcome: Cross-functional accountability and consistent governance.

3. **Likelihood & Impact Assessment:**
   Use scenario planning and expert judgment due to limited historical data.
   Outcome: Realistic evaluation of emerging and low-frequency, high-impact events.

4. **Risk Treatment Approaches:**

   - *Avoidance:* Decline high-risk AI use cases.
   - *Mitigation:* Apply AI-specific controls (adversarial defenses, monitoring, governance).
   - *Transfer:* Evaluate insurability of AI-related risks.
   - *Acceptance:* Document rationale and monitor conditions.

5. **Risk Governance Alignment:**
   Update risk registers, risk appetite statements, and internal controls to include AI.
   Outcome: Accountability at the board and executive levels.

Board Implication: AI risk cannot be contained to IT or security. It must be governed as an enterprise-wide strategic risk with clear ownership, reporting cadence, and resource allocation.

3. Continuous Monitoring of AI Models and Performance

AI systems degrade over time and must be continuously monitored to maintain effectiveness. This monitoring cannot be optional as it is an operational necessity.

Key Monitoring Priorities

- **Model & Data Drift Detection:**
  Monitor changes in prediction accuracy, confidence scores, and data distributions.
- **Adversarial Behavior Detection:**
  Track anomalies, perturbation patterns, and unexplained misclassifications.
- **Operational Health Metrics:**
  Observe model latency, throughput, resource utilization, and data pipeline stability.
- **Bias & Fairness Indicators:**
  Monitor performance differences across demographic or organizational groups.

Monitoring Infrastructure Recommendations

- Role-based dashboards for executives, security leaders, and data scientists
- Automated alerting for deviation from baselines
- Continuous retraining pipelines (MLOps)
- Canary deployments for updated models
- Integration with SIEM/SOAR platforms

Board Implication: AI systems must be treated like living assets requiring real-time telemetry and maintenance. Without continuous monitoring, AI can silently fail in ways humans cannot detect until damage occurs.

4. Establishing Measurable Key Risk Indicators (KRIs) for AI

KRIs translate complex AI behavior into quantifiable signals executives can track. They serve as the organization's early warning system.

Essential KRI Categories

1. **Model Performance KRIs**

   - Rolling accuracy/precision/recall trends
   - False positive/false negative thresholds
   - Confidence score distribution shifts

2. **Threat Adaptability KRIs**

   - Detection rates for novel attack types
   - AI evasion success in internal red team tests
   - Time-to-detect and time-to-respond for AI-flagged incidents

3. **Ethical & Compliance KRIs**

   - Bias and fairness drift across user groups
   - PII anonymization success rates
   - Auditability of AI decisions

4. **Operational Health KRIs**

   - GPU/CPU utilization trends
   - Data pipeline latency/error rates
   - Retraining frequency and stability

Board Implication: KRIs give leadership a direct view into AI health, risk exposure, and compliance posture. They transform AI from an opaque capability into an accountable and measurable risk domain.

5. Developing Incident Response Plans for AI-Related Failures

AI failures require response processes that differ fundamentally from traditional cybersecurity incidents.

AI-Specific Failure Types

- Model drift causing degraded accuracy
- Data poisoning compromising model behavior
- Adversarial inputs triggering misclassifications
- Data pipeline failures creating incomplete or corrupted input
- Infrastructure overload reducing model responsiveness

AI-Specific Incident Response Requirements

1. **AI-Failure Detection Triggers**

   - KRI threshold breaches
   - Sudden shifts in prediction patterns
   - Analyst-identified anomalies

2. **Cross-Functional Investigation Teams**

   - SOC analysts
   - MLOps engineers
   - Data scientists
   - Legal/compliance as needed

3. **Containment Actions**

   - Disable automated responses
   - Move model to monitoring-only mode
   - Fallback to rule-based controls or human review

4. **Eradication & Recovery**

   - Retrain model with verified data
   - Remove poisoned samples

- Validate with simulation and red teaming
- Canary deploy restored models

5. **Documentation & Reporting**

   - Preserve model states for forensic review
   - Record incident timelines and decisions
   - Communicate to regulators or customers when appropriate

6. **Exercises & Simulations**

   - AI-specific tabletop scenarios
   - Adversarial stress tests
   - Model failure drills

Board Implication: Treat AI failures as high-priority security incidents. Preparedness reduces downtime, prevents cascading failures, and demonstrates responsible governance in an AI-augmented enterprise.

Conclusion for the Board

AI will significantly enhance organizational security capabilities, but only if its risks are governed with the same rigor applied to financial controls, cyber defenses, and regulatory compliance. Effective AI risk management must be integrated into ERM, continuously monitored, quantified through KRIs, and supported by AI-specific incident response plans.

Boards must ensure:

- Clear AI risk ownership
- Ongoing visibility through AI KPIs/KRIs
- Adequate investment in MLOps and monitoring infrastructure
- Ethical and compliant deployment standards
- Preparedness for AI-specific failures

AI is a force multiplier; but without disciplined governance, it can become an unmonitored systemic risk.

Conclusion: From AI Adoption to AI Risk Mastery

Artificial intelligence fundamentally alters the risk equation in cybersecurity by introducing systems that learn, adapt, and operate beyond deterministic control. As this chapter has demonstrated, AI-related risks are not confined to technical failures but extend across governance, compliance, operations, ethics, and resilience. Model integrity, data poisoning, adversarial manipulation, drift, and over-automation represent persistent risk conditions rather than isolated events. Managing these risks therefore requires sustained executive attention and structured oversight, not one-time controls or reactive remediation.

Strategic AI risk management begins with recognizing AI systems as high-impact assets whose behavior must be governed across their full lifecycle. Integrating AI risks into enterprise risk management ensures visibility at the executive and board levels and enables informed decisions about risk appetite, investment, and accountability. Continuous monitoring transforms AI from a static deployment into a managed capability, while measurable Key Risk Indicators provide early warning signals that prevent silent degradation and compounding failures. These mechanisms collectively shift AI security from assumption-based trust to evidence-based confidence.

Equally important is preparing for AI-specific failures with intentional incident response planning. AI incidents demand specialized detection, containment, forensic analysis, and recovery processes that acknowledge probabilistic behavior and opaque decision-making. Organizations that plan for AI failure preserve operational continuity and trust, while those that assume infallibility risk cascading disruption. Human oversight remains the stabilizing force in this model, ensuring that automation enhances rather than erodes resilience.

Ultimately, effective AI risk management is not a barrier to innovation but its enabler. By embedding governance, monitoring, and response into AI security operations, leaders create the conditions under which AI can be deployed responsibly, confidently, and at scale. The organizations that succeed will be those that treat AI risk as a strategic discipline, aligning technical safeguards with executive oversight and human judgment. In doing so, AI becomes not a source of uncertainty, but a controlled and durable advantage in an increasingly complex threat environment.

Chapter 5: AI-Augmented Security Operations Center (SOC)

Artificial Intelligence is reshaping the Security Operations Center by shifting the SOC from reactive alert handling to intelligence-led operations. While traditional automation reduces manual effort, AI changes the quality of decision-making by improving situational awareness, correlating weak signals, enriching investigative context, and enabling proactive discovery of novel threats. This chapter explains how AI augments the SOC across alert triage and prioritization, IOC enrichment, threat hunting, predictive analysis, and human-AI teaming, where machines provide scale and humans preserve judgment and accountability. It also establishes why ethical and trustworthy AI must be treated as an operational requirement, not a policy add-on, because SOC AI influences decisions that affect business continuity, user experience, and compliance exposure. The objective is to equip security leaders with a practical framework for deploying AI in the SOC in a way that measurably improves outcomes while maintaining transparency, fairness, and control.

Transforming the SOC with AI Beyond Automation

The Security Operations Center (SOC) has long served as the frontline defense where analysts monitor, detect, and respond to cyber threats under relentless time pressure. Modern enterprises now generate volumes of telemetry and alerts that exceed what traditional SOC models were designed to absorb, and attackers increasingly operate with stealth and multi-stage techniques that evade static rules. Artificial Intelligence (AI) transforms the SOC not merely by accelerating tasks, but by improving how the SOC understands risk, prioritizes activity, and composes decisions. "Beyond automation" matters because automation executes predefined steps, while AI can learn, correlate, and infer

125

context across complex environments. In the SOC, this enables a shift from reactive queue management to intelligence-led operations. The goal is not to replace analysts but to elevate their work by reducing noise and increasing the quality of investigative context.

AI strengthens situational awareness by ingesting and correlating signals across endpoints, network sensors, identity platforms, cloud services, applications, and external threat intelligence. Instead of analysts manually pivoting between disconnected tools to reconstruct an incident, AI can surface relationships among events that appear benign when viewed in isolation. For example, an AI system can connect anomalous login behavior to endpoint process execution, unusual outbound traffic, and active exploitation chatter relevant to the impacted technology stack. This integrated view shortens the time required to understand what is happening and what assets are at risk. It also improves decision quality by prioritizing signal over volume and providing explainable context. Analysts remain responsible for judgment, but they act with a clearer narrative of threat progression.

AI also changes threat hunting by expanding what can be discovered beyond known signatures and static indicators. Traditional threat hunting often begins with a human hypothesis and requires time-consuming log searches, which limits coverage and makes hunts episodic rather than continuous. AI can surface subtle anomalies and suspicious sequences of actions that deviate from learned baselines, even when no known indicators exist. It can also generate likely hypotheses by mapping observed behaviors to attacker tactics, techniques, and procedures, helping hunters focus on the most probable paths. This enables the SOC to identify Indicators of Attack rather than relying solely on Indicators of Compromise after damage is underway. As a result, dwell time decreases because suspicious behavior is detected earlier and investigated faster. This is augmentation in practice: AI expands the search space while humans apply reasoning and operational context.

AI-driven SOC transformation also streamlines the detection-to-response workflow by reducing alert fatigue and accelerating triage decisions. Alert queues are often dominated by false positives, repetitive noise, or isolated events that lack business context, which degrades attention and increases risk. AI improves prioritization by weighting alerts using asset criticality, identity risk, behavioral deviation, correlated evidence, and threat intelligence alignment. It can also enrich alerts automatically, gathering relevant system history, user context, vulnerability exposure, and related events into a single investigative package. This reduces time spent collecting context and increases the speed and consistency of initial response. The outcome is not simply faster detection, but faster understanding and more precise containment decisions. Over time, the SOC becomes less reactive and more deliberate, operating with higher signal and lower volatility.

A practical example of this human-AI collaboration occurs when AI identifies weak signals that rule-based systems overlook. An AI model may detect persistent encrypted outbound connections from a server that normally does not generate such traffic, then correlate the activity with recent system changes, credential events, and the server's role in sensitive workflows. Individually these signals may appear explainable, but together they can indicate command-and-control activity or staged exfiltration. The analyst begins with a curated lead rather than guessing where to look, and the AI provides supporting evidence that accelerates hypothesis testing. The analyst then determines whether the pattern reflects misconfiguration, a legitimate exception, or malicious behavior requiring containment. The AI provides scale and correlation, while the human provides judgment and accountability. This combination improves outcomes without creating blind reliance on automation.

AI insights generated in the SOC can also feed proactive defense programs such as vulnerability management, identity hardening, and targeted user training. If AI repeatedly identifies phishing campaigns targeting specific roles, those patterns can guide more focused awareness interventions and policy adjustments. If AI-driven threat intelligence detects active exploitation of a technology stack present in the organization, it can elevate remediation priority and trigger validation scans. This converts SOC operations into a broader intelligence function that influences risk reduction before incidents occur. The result is a more predictive security posture that integrates detection signals into prevention strategy. While the SOC remains responsible for response, it has become a producer of actionable intelligence across the enterprise. This is one of the clearest indicators that AI has moved the SOC beyond automation.

Optimizing Alert Triage and Prioritization with AI

The volume of alerts produced by modern security tools can overwhelm even mature SOCs and create persistent alert fatigue. When analysts are forced to evaluate hundreds or thousands of alerts daily, critical incidents can be overlooked, misclassified, or delayed in investigation. This problem is typically systemic, rooted in rule-based alerting that lacks sufficient context rather than analyst capability. AI addresses the issue by ranking alerts based on correlated evidence and probable impact rather than treating them as independent events. Instead of presenting a flat queue, AI can reduce noise by suppressing known-benign patterns and elevating alerts that appear connected to meaningful attack chains. This improves both response speed and decision quality. The objective is to give analysts fewer, higher-fidelity decisions.

AI-driven triage is powered by correlation across identity, endpoint, network, cloud, and application telemetry. A failed login attempt, for example, can be evaluated alongside geo-velocity anomalies, unusual device posture, privileged access context, and post-authentication behaviors. AI can incorporate asset criticality, vulnerability exposure, lateral movement signals, and known attacker infrastructure to refine severity. This correlation enables detection of multi-stage attacks in which early steps appear harmless when viewed alone. It also reduces the time analysts spend assembling context from multiple consoles and data stores. When correlation is reliable, AI can provide a narrative view of the incident rather than a collection of alerts. Analysts can then validate the narrative and move quickly to containment decisions.

Machine learning models further improve prioritization by learning patterns associated with true incidents and recurring false positives in the organization's environment. Supervised learning can capture labeled outcomes from prior investigations, while anomaly detection can surface novel behaviors outside learned baselines. This supports detection of threats without known signatures, including low-and-slow intrusion activity and living-off-the-land techniques. Analyst feedback is essential because it refines what "normal" means in a specific environment and prevents the model from amplifying noise. Over time, this feedback loop increases alert fidelity and reduces wasted effort. The SOC becomes faster not because it works harder, but because it works on the right problems first. This is a meaningful operational advantage in environments where attacker speed is a defining factor.

AI-augmented triage also improves the analyst experience by providing pre-assembled investigative context with each prioritized alert. A high-severity item should include impacted identities, involved assets, correlated events, relevant log excerpts, external intelligence alignment, and suspected tactics and techniques. This allows analysts to move directly into validation and response instead of spending the first phase of investigation gathering evidence. AI can also propose response options based on incident type and confidence, while preserving human approval for disruptive actions. This accelerates containment while reducing unintended operational harm. It also reduces burnout by limiting repetitive queue-chasing and lowering cognitive overload. The SOC becomes more sustainable and more resilient under surge conditions.

Enriching Indicators of Compromise IOCs with AI

Indicators of Compromise (IOCs) such as malicious IPs, file hashes, domains, or registry artifacts are valuable signals, but they rarely provide enough context to drive a confident response decision on their own. Analysts typically enrich IOCs by querying SIEM data, searching endpoint telemetry, consulting asset inventories, and checking multiple threat intelligence sources. This work is essential, but it is time-consuming and becomes a bottleneck during complex incidents with many artifacts. AI accelerates enrichment by attaching operational meaning, historical context, and environmental exposure to raw indicators. It transforms IOCs from isolated "clues" into decision-ready intelligence packages. The goal is to shorten the time from "indicator observed" to "scope understood" and "response justified." In high-tempo incidents, that time recovery can prevent escalation.

AI enrichment adds external context by consulting threat intelligence and mapping indicators to known infrastructure, campaigns, or malware families where possible. Rather than merely labeling an IP as suspicious, AI can identify whether it aligns with phishing infrastructure, ransomware command-and-control, scanning activity, or credential harvesting. This classification supports faster scoping and helps analysts anticipate likely next steps. AI can also link indicators to attacker tactics and techniques, enabling more consistent investigation and response workflows. When the specific indicator is new, AI can still infer risk through infrastructure similarity, behavioral cues, and historical campaign features. This reduces dependence on any single feed and strengthens resilience against novel infrastructure rotation. The analyst receives context that is actionable rather than generic.

Internal context is often the decisive factor, and AI excels at correlating indicators with enterprise-specific risk. If a hash appears on an endpoint, AI can identify that endpoint's role, recent activity changes, exposure to sensitive data, and relationship to privileged accounts. If a domain appears in DNS or proxy logs, AI can show which hosts contacted it, when the activity began, whether the destination is novel for that segment, and what data movement patterns followed. This prevents underestimating indicators tied to critical assets and prevents overreaction when indicators appear in benign contexts. Asset criticality and identity privilege often determine severity more than the indicator alone, and AI can attach those factors consistently. The result is prioritization grounded in the organization's environment, not only in external reputation scoring. That improves both accuracy and executive defensibility.

AI can also reconstruct attack progression by linking multiple IOCs into an incident timeline that reflects the attacker's sequence of actions. A single campaign may include phishing entry points, payload downloads, persistence artifacts, lateral movement indicators, and exfiltration destinations, each visible in different tools. AI can stitch these events into a coherent chain, showing scope, suspected objectives, and affected systems. This supports more accurate containment and eradication because teams understand what occurred and what remains at risk. It also improves reporting to leadership because the narrative is clearer and evidence based. Over time, these reconstructions strengthen detection engineering by highlighting weak points in telemetry and control coverage. The SOC becomes better at learning from incidents rather than simply closing tickets.

AI-Powered Threat Hunting and Predictive Analysis

AI-powered threat hunting advances the SOC from reactive alert handling to proactive discovery of hidden adversary behavior. Traditional threat hunting depends heavily on human intuition and manual log analysis, which limits scale and makes hunts intermittent. AI can continuously analyze network traffic, endpoint behavior, identity signals, and application activity to identify deviations from learned baselines. These anomalies may be subtle, such as unusual access patterns, abnormal process chains, or unexpected outbound communications that do not match known signatures. AI can then surface these patterns as prioritized leads with supporting context rather than producing raw noise. This reduces the search space for hunters and increases the probability of finding meaningful threats earlier. The outcome is shorter attacker dwell time and improved detection of stealthy intrusion patterns.

AI is particularly effective against advanced persistent threats and zero-day exploitation because these threats often rely on behavior, not obvious malware signatures. Unsupervised learning methods can identify multi-step attack sequences where each individual step appears ambiguous, but the combined chain indicates intrusion. AI can flag unusual privilege escalation paths, lateral movement behavior, suspicious credential use, and anomalous access to sensitive data stores. It can also detect abnormal usage of legitimate administrative tools commonly used for living-off-the-land techniques, such as unusual execution cadence, atypical command patterns, or unexpected parent-child process relationships. These detections are difficult for static rules because "malicious" is contextual and often resembles legitimate operations. AI improves this by learning what "normal" looks like in the organization's actual workflows and highlighting meaningful deviations. Human validation remains essential, but AI increases the probability that human time is spent on the right anomalies.

Predictive analysis extends hunting by anticipating likely attacker paths and enabling preventative action before exploitation succeeds. AI can combine vulnerability exposure, asset criticality, network topology, identity privileges, and external threat intelligence to identify which systems are most likely to be targeted and how attackers might traverse the environment. This supports risk-aligned patch prioritization, segmentation decisions, and control hardening focused on the most plausible attack routes. AI can also forecast which emerging tactics are gaining traction in comparable industries, allowing the SOC to adjust detection coverage and readiness. Predictive outputs should be treated as decision support rather than certainty, but they improve planning under uncertainty. When validated through adversary simulation and red team exercises, predictive insights become operationally credible. The SOC evolves into an intelligence hub that supports both response and prevention.

AI can also help leadership understand business impact by modeling the consequences of compromise in systems with operational dependencies. By mapping how critical processes rely on specific applications, data stores, and identity services, AI can highlight where failure would cascade. This supports more informed prioritization of resilience investments and incident response preparation. It also improves executive communication because security teams can explain risk in operational and financial terms rather than only technical severity. Predictive analysis does not replace governance, but it strengthens governance decisions with evidence and structured modeling. When combined with continuous monitoring, it supports a more mature security posture. The SOC becomes not just a detection center, but a contributor to enterprise resilience strategy.

Human-AI Teaming for Enhanced SOC Performance

A mature SOC does not aim to remove humans from security decisions, but to design workflows where AI and analysts complement one another. AI provides scale through correlation, enrichment, prioritization, and rapidly gathering evidence, while analysts contribute judgment, context, creativity, and accountability. This model improves performance because analysts are no longer trapped in repetitive triage and can focus on complex investigations, detection engineering, and proactive hunting. AI acts as a co-pilot that reduces cognitive load and accelerates the early stages of incident understanding. The analyst remains the decision authority for high-impact actions and exceptions where business context matters. When implemented correctly, human-AI teaming improves speed without sacrificing governance. The SOC becomes both more effective and more sustainable.

In incident response workflows, AI can assemble timelines, link related events, and highlight the most likely attack narrative before the analyst begins deep investigation. This compresses the period where time is often lost to manual pivoting across tools and helps reduce mean time to triage and contain. For common incident types, AI can execute standardized investigative steps, such as gathering endpoint telemetry, correlating identity and network events, and enriching artifacts with intelligence context. Where appropriate, AI can recommend containment actions, but human approval remains essential to prevent unnecessary disruption. This approach improves consistency across shifts and experience levels because response quality becomes less dependent on individual bandwidth. It also supports surge operations by allowing the SOC to maintain discipline under volume spikes. AI increases throughput, while humans preserve correctness and accountability.

Human-AI teaming becomes durable only when it includes a structured feedback loop that improves model performance over time. Analysts must be able to label outcomes, confirm true positives, mark false positives, and record reasoning that helps refine prioritization and detection logic. This feedback should flow into MLOps processes that update rules, retrain models, and validate changes through testing and controlled deployment. Without feedback, AI performance stagnates and drift accumulates, creating operational risk and eroding trust. With feedback, AI becomes increasingly tailored to the organization's environment, reducing noise and improving fidelity. This transforms AI from a static tool into a continuously improving capability. The SOC improves through operations, not just through periodic platform upgrades.

Trust calibration is also a critical element of human-AI teaming because both over-trust and under-trust create failure modes. Over-trust leads to rubber-stamping AI outputs and can result in missed threats or disruptive actions taken on incorrect confidence. Under-trust forces analysts to re-do AI work manually, eliminating the efficiency benefits and creating resentment toward the tooling. The SOC must define where AI provides decision support versus where AI is permitted to execute actions, and those boundaries should align to impact and confidence. High-impact actions such as account lockouts, host isolation, or production-blocking controls should typically require human confirmation or staged automation. Establishing these guardrails improves both governance and operational reliability. It also makes performance more predictable and defensible.

AI-enabled SOC performance also depends on workforce evolution, because analysts must be equipped to interpret AI outputs and recognize model limitations. Analysts do not need to become data scientists, but they do need baseline literacy in concepts such as false positives, drift, confidence, and explainability. SOC leaders also need supporting roles, including detection engineers, data pipeline owners, and MLOps partners who can sustain model health and telemetry quality. Training and playbook updates must be part of adoption, so the SOC remains resilient when models change, environments shift, or adversaries adapt. This skill evolution protects the organization from the operational risk of AI dependency. It also improves retention by elevating analyst work toward higher-value activities. Human-AI teaming succeeds when people and process mature alongside the technology.

Ultimately, the AI-augmented SOC is best understood as an operating model in which intelligence is produced continuously and decisions are made faster with stronger evidence. AI expands the SOC's capacity to correlate weak signals, enrich investigations, and proactively hunt for threats that evade traditional controls. Humans ensure that decisions remain aligned with business context, ethics, and accountability, especially when actions are disruptive or ambiguous. When AI is integrated with clear guardrails and a learning loop, the SOC becomes more predictive, more efficient, and more resilient. This is the practical promise of AI in security operations: better outcomes at scale without surrendering control. The future SOC is not automated security, but orchestrated security driven by human judgment and machine speed. That is how organizations defend effectively in an environment where adversaries evolve continuously.

Ethical and Trustworthy AI in SOC Operations

Ethical and trustworthy AI is not an abstract concern in the Security Operations Center, because SOC systems routinely influence decisions that affect access, productivity, investigations, and sometimes employment outcomes. AI-driven triage, enrichment, and response recommendations can shape how incidents are interpreted and how people are treated, especially when identity data and behavioral analytics are involved. Trustworthy SOC AI therefore requires deliberate governance that balances security effectiveness with fairness, accountability, and operational safety. The SOC must be able to justify why an AI system elevated an alert, recommended an action, or inferred risk about a user or device. This is essential for internal legitimacy and for external defensibility under regulatory and legal scrutiny. Ethical AI in SOC operations is best treated as a security requirement because it reduces error-driven harm and strengthens organizational resilience.

A core ethical risk in SOC AI is bias, which can appear when models learn from incomplete or unrepresentative training data or when proxies for sensitive attributes emerge in features. In practice, this can create uneven false positive rates across departments, locations, job functions, or user privilege tiers, leading to disproportionate investigations or disruptions. Bias can also appear when the SOC's historical labels reflect prior operational assumptions rather than objective ground truth, causing models to learn "who gets investigated" rather than "what is malicious." Trustworthy SOC AI requires continuous fairness testing, including measuring false positive and false negative deltas across defined groups and monitoring for drift over time. When disparities appear, the SOC should treat them as operational risk, not as a public relations issue, because uneven accuracy reduces security reliability. Mitigation often involves feature review, improved labeling practices, broader training data, and formal review of high-impact automation boundaries.

Explainability and auditability are equally important because SOC decisions often require justification, especially when actions are disruptive. Deep models may provide strong detection performance but offer limited transparency into why a decision was made, which can undermine analyst trust and complicate incident review. A trustworthy SOC AI system should provide interpretable evidence trails such as contributing signals, correlated events, relevant baselines, and confidence context rather than delivering a single opaque score. Auditability also supports compliance expectations, internal controls, and post-incident learning, because teams can review how decisions were reached and refine the system responsibly. Where full explainability is not feasible, the SOC should establish compensating controls such as stricter human review thresholds, additional validation steps, and more conservative automation policies. The guiding principle is that the more disruptive the outcome, the higher the transparency and oversight required.

Operational trust also depends on strong data governance, because SOC AI systems rely on sensitive telemetry, identity data, and sometimes personal information. Data minimization should be applied intentionally so models receive only what is necessary for security outcomes, rather than collecting broadly because AI "can use it." Retention policies must be aligned to security needs, legal constraints, and privacy expectations, particularly for user behavior analytics and communications metadata. Access controls for training data, feature stores, and model outputs must be treated as high-value security domains, because compromises can expose sensitive information and enable model manipulation. The SOC should also monitor data leakage risk through model outputs, including the inadvertent exposure of sensitive attributes in dashboards, case notes, or automated summaries.

Strong privacy practice supports trust internally and reduces the blast radius of compromise externally.

Trustworthy AI in the SOC also requires explicit guardrails for automation, because "correctness" is not only a model property but a business-impact decision. The SOC should classify actions by impact and require human approval for high-impact steps such as account lockouts, endpoint isolation, production blocking, or HR-sensitive escalation. Lower-impact steps such as enrichment, evidence gathering, and case routing can be automated more aggressively, provided they remain observable and reversible. Confidence thresholds should be calibrated to the organization's risk appetite, and exceptions must be handled consistently so that urgent incidents do not bypass safeguards in ways that become routine. This is where governance becomes practical: it defines what AI may recommend, what AI may execute, and where humans must intervene. Clear boundaries preserve speed while preventing avoidable harm and building long-term trust in AI-enabled operations.

Finally, ethical SOC AI requires continuous assurance through monitoring, testing, and accountability mechanisms that make trust measurable rather than aspirational. This includes monitoring model drift, tracking error patterns, running adversarial evaluations, and performing periodic audits of decisions influenced by AI. It also includes an incident response process for AI failure modes, where model errors, bias incidents, or suspicious manipulations are treated as operational events with clear escalation and remediation steps. The SOC should maintain a learning loop that captures analyst feedback, model performance outcomes, and governance decisions, ensuring the system improves without becoming unstable. When accountability is defined and enforced, analysts trust the tooling more, leaders gain defensible oversight, and the organization reduces risk created by opaque automation. Ethical and trustworthy AI

139

therefore becomes a force multiplier, strengthening the SOC's credibility and resilience while improving security performance.

Board-Ready Brief

Executive Overview

Security Operations Centers are strained by growing alert volume, fragmented telemetry, and adversaries who move faster and conceal activity within normal workflows. Artificial Intelligence improves SOC performance by reducing noise, correlating weak signals into meaningful narratives, accelerating triage and investigation, and enabling proactive threat discovery. The value proposition is not automation alone, but augmentation, where AI improves decision quality and analyst leverage while humans retain accountability for high-impact actions. This chapter explains how AI transforms SOC operations across situational awareness, alert prioritization, IOC enrichment, threat hunting, predictive analysis, and human-AI teaming. It also establishes why ethical and trustworthy AI is a governance requirement, not a policy add-on, because AI-driven SOC decisions affect people, business operations, and compliance exposure. For executive leadership and boards, the focus should be measurable operational outcomes paired with disciplined oversight and guardrails.

1. Transforming the SOC Beyond Automation

Traditional SOC workflows rely heavily on rule-based alerts and manual correlation across tools, which are increasingly mismatched to modern enterprise complexity. AI enables the SOC to learn normal behavior, detect deviations, and connect events across identity, endpoint, network, cloud, and application data. This improves situational awareness by turning disconnected telemetry into context-rich narratives that analysts can validate and act on quickly. The SOC shifts from reactive alert handling toward intelligence-led operations, where higher fidelity insights guide response and prevention. The primary operational advantage is increased throughput and improved decision consistency without requiring linear staffing growth. Leadership should view AI as a capability multiplier that must be governed like critical control, not treated like a plug-in feature.

2. Optimizing Alert Triage and Prioritization

Alert fatigue is a persistent SOC risk because it increases the probability of missed incidents and delays containment. AI improves alert triage by correlating evidence across data sources and prioritizing alerts by probable impact rather than by individual triggers. It can suppress repetitive benign patterns, elevate signals that align with meaningful attack chains, and deliver alerts with pre-assembled context. This reduces time wasted on manual evidence gathering and increases the speed and quality of triage decisions. AI prioritization becomes stronger when it learns from analyst outcomes, improving fidelity over time. The board-level outcome is reduced mean time to triage and higher confidence that critical threats are acted upon quickly.

3. Enriching Indicators of Compromise

Raw Indicators of Compromise often require significant manual effort to interpret and scope, especially during complex incidents. AI automates enrichment by attaching external intelligence, internal telemetry correlations, asset criticality, identity context, and behavioral signals to indicators. It can also link multiple related indicators into an attack timeline, revealing progression, scope, and likely objectives. This improves containment and eradication because teams can act with a clearer picture of what has occurred and what remains at risk. It also improves executive reporting by enabling evidence-based narratives rather than fragmented artifact lists. The measurable outcome is reduced time to understand scope and improved accuracy in response decisions.

4. AI-Powered Threat Hunting and Predictive Analysis

AI strengthens proactive threat hunting by continuously identifying anomalies and suspicious sequences that do not match known signatures. This is particularly useful against advanced persistent threats, zero-day exploitation, and living-off-the-land techniques where behavior is the primary signal. Predictive analysis extends these benefits by forecasting likely attack paths, highlighting exposed assets most likely to be targeted, and informing risk-aligned prioritization of defensive actions. These capabilities shift the SOC toward prevention by enabling control hardening before exploitation succeeds. Leadership should expect these capabilities to be validated through adversary simulation and measured through reduced dwell time and improved detection of novel threats. The SOC has become an intelligence hub supporting both response and strategic risk reduction.

5. Human-AI Teaming as the Operating Model

The most effective SOC is built on human-AI teaming, not human replacement. AI performs scale tasks such as correlation, enrichment, prioritization, and repetitive investigative steps, while analysts provide judgment, context, and accountability. This requires workflow redesign so AI outputs are actionable and auditable, and analysts can provide structured feedback to improve model performance over time. High-impact actions should remain gated by human review to prevent disruption from confident errors. Workforce impacts are expected, including increased demand for data-literate analysts, detection engineering skills, and MLOps partnership for sustained model health. The board-level message is that AI improves SOC performance when governance, feedback loops, and training are treated as core adoption requirements.

6. Ethical and Trustworthy AI in SOC Operations

Ethical and trustworthy AI is a SOC requirement because AI outputs influence actions that can disrupt business operations and affect individuals. Bias can appear in unequal false positive rates across roles, locations, or departments, creating operational risk and legitimacy challenges. Explainability and auditability matter because SOC actions must be justified during incident review, compliance inquiries, or internal disputes. Data governance is essential because SOC AI process's sensitive identity and behavioral information, requiring minimization, access control, retention discipline, and monitoring for leakage through outputs. Guardrails must classify what AI may recommend versus execution, with human approval reserved for high-impact actions such as lockouts, isolation, or HR-sensitive escalation. Continuous assurance—monitoring drift, testing adversarial resilience, and auditing AI-influenced decisions—makes trust measurable and defensible.

Oversight Questions for Executives and Boards

- Which SOC decisions are AI-influenced versus AI-executed, and where is human approval mandatory?

- What measurable outcomes demonstrate AI is improving security operations: false positive reduction, mean time to triage, dwell time reduction, containment speed, or incident cost reduction?

- How are AI models monitored for drift, adversarial manipulation, bias, and error clustering?

- What is the operational fallback when models degrade or produce confident but incorrect conclusions?

- How are we governing data used for SOC AI, including retention, access, privacy constraints, and auditability?

- How are we preventing over-reliance while preserving analyst expertise and decision accountability?

Conclusion for the Board

AI can materially improve SOC effectiveness by increasing signal quality, accelerating investigations, and enabling proactive threat discovery, but it also introduces operational dependencies that must be governed. Successful AI-augmented SOC adoption requires integration across telemetry sources, measurable performance objectives, human-AI teaming workflows, and explicit guardrails for high-impact actions. Ethical and trustworthy AI is not optional because bias, opacity, and poor data governance create both security risk and organizational harm. Boards should demand visibility into AI performance metrics and assurance controls, ensure accountability for AI risk ownership, and confirm that the SOC maintains resilient fallbacks when AI systems degrade. When governed well, AI becomes a force multiplier that strengthens resilience without sacrificing control.

Conclusion: The AI-Augmented SOC as a Strategic Asset

An AI-augmented SOC succeeds when it is built as an operating model, not a tool upgrade. AI strengthens SOC performance by improving situational awareness, reducing alert fatigue through better prioritization, accelerating investigations through automated enrichment, and enabling proactive threat hunting that detects behavior beyond known signatures. These advantages compound when human-AI teaming is designed intentionally, with analysts providing judgment and feedback while AI delivers scale, correlation, and speed. Trust and ethics are central to durability because bias, opacity, and weak data governance can create operational harm, erode analyst confidence, and introduce compliance risk even when detection performance appears strong. The SOC must therefore implement guardrails for automation, ensure auditability of AI-influenced decisions, and maintain continuous assurance through monitoring, testing, and documented accountability. When these elements mature together, AI becomes a force multiplier that improves resilience and decision quality without surrendering control, enabling the SOC to defend more effectively in an environment where adversaries evolve continuously.

Chapter 6: AI in Incident Response (IR)

Incident Response is the most consequential function within a cybersecurity program because it determines how an organization performs when control fails and adversaries gain a foothold. Detection alone does not protect the enterprise; protection is realized through swift containment, informed decision-making, and disciplined recovery. As cyber incidents grow faster, more automated, and more interconnected, traditional response models increasingly struggle to keep pace.

Artificial Intelligence fundamentally alters Incident Response by accelerating understanding, improving response consistency, and enabling organizations to act earlier in the attack lifecycle. Rather than replacing human responders, AI augments their ability to interpret complex signals, scope impact accurately, and execute containment with confidence. This chapter explores how AI transforms Incident Response across investigation, decision support, automation, reporting, and organizational learning.

The discussion emphasizes not only technical acceleration, but also governance, trust, and operational discipline. AI-enabled Incident Response succeeds only when speed is balanced with accountability, explainability, and human oversight. When implemented as a governed capability, AI allows organizations to reduce attacker dwell time, limit operational disruption, and convert incidents into measurable improvements in resilience.

Accelerating Incident Response with AI Capabilities

Incident Response is where cybersecurity becomes operational resilience, because every minute of uncertainty increases the likelihood of spread, disruption, and loss. The core phases of Incident Response remain consistent: detect, analyze, contain, eradicate, and recover, followed by lessons learned and control improvement. What has changed is the speed, scale, and complexity of modern incidents, especially in hybrid environments spanning endpoints, identity providers, cloud services, and third party platforms. Human responders still lead incident command and judgment, but they are increasingly constrained by data fragmentation and time intensive correlation work. Artificial Intelligence accelerates Incident Response by compressing the distance between an alert and a verified incident narrative that supports decisive action. When AI is governed correctly, the outcome is faster containment, reduced dwell time, and more reliable recovery.

The acceleration begins in the earliest moments of an event, where organizations must decide whether a signal is noise or the first indicator of material compromise. Traditional alert pipelines generate volume, but not always clarity, and analysts often spend early hours validating basic facts across multiple tools. AI can ingest alerts and telemetry at scale and perform rapid correlation across endpoint behavior, identity events, network signals, and threat intelligence context. This turns isolated alerts into incident hypotheses that reflect likely attacker intent and probable next steps. The value is not only speed but also triage quality, because prioritization improves when risk scoring reflects asset criticality, privilege level, and blast radius potential. A strong AI assisted triage layer reduces analyst overload while increasing the probability that high severity incidents receive immediate attention.

AI also changes the practical rhythm of Incident Response by shifting responders away from manual evidence collection toward evidence validation and decision-making. Investigation often stalls because responders must gather logs, normalize formats, align timestamps, and reconstruct sequences of actions across disconnected sources. AI driven investigation can automate data retrieval and build timelines that are continuously updated as new evidence arrives. This enables earlier containment decisions based on a coherent view of scope, rather than waiting for complete certainty while the adversary continues to operate. The result is a more proactive posture where responders can act on high confidence signals and refine containment as additional evidence confirms the full picture. This is the difference between chasing alerts and managing an incident with disciplined tempo.

AI Enabled Detection, Triage, and Scoping

Triage and scoping are the leverage points of Incident Response, because they determine how quickly the organization moves from suspicion to action. AI improves triage by correlating low fidelity signals into higher confidence narratives that better represent real attacker behavior. A single identity anomaly may be harmless on its own but paired with endpoint process anomalies and unusual outbound traffic, it can become a compelling incident hypothesis. AI can score this hypothesis using contextual factors such as the sensitivity of the targeted system, the privilege level of the account, the time of activity relative to baseline, and whether related indicators appear elsewhere in the environment. This enables responders to focus on a smaller set of high value investigations rather than a long queue of isolated alerts. The practical effect is reduced time to first meaningful action and improved consistency across shifts and teams.

Scoping is where organizations either contain effectively or allow the adversary to widen the compromise footprint. AI accelerates scoping by rapidly identifying related assets, identities, and network paths that are likely involved in the same incident cluster. For example, if an endpoint shows evidence of credential dumping, AI can immediately expand analysis to other systems accessed by that identity, other identities used on the same device, and systems sharing authentication dependencies. If the incident involves possible data exfiltration, AI can focus on high volume transfers, unusual destinations, and traffic patterns that indicate staging behavior. This scoping capability is strongest when AI is integrated with asset inventories and business service maps, because impact is not defined by the number of systems but by the criticality of what they support. A mature program uses AI to move quickly from tactical scoping to service oriented scoping that reflects actual business risk.

AI supported triage and scoping must remain disciplined, because acceleration without governance can create a different risk: fast decisions based on weak evidence. The program should define which AI outputs are advisory and which are treated as high confidence triggers for response actions. This includes setting confidence thresholds, requiring corroboration from multiple independent signals for disruptive actions, and ensuring incident command retains decision authority. AI should be expected to justify risk scores through explainable factors such as correlated indicators and deviations from baselines, not only opaque model outputs. When AI outputs remain auditable and evidence linked, responders can move faster without losing defensibility. This is essential for both operational trust and post incident accountability.

Automated Investigation and Root Cause Analysis

Investigation is traditionally the most labor-intensive phase of Incident Response, because responders must unify evidence across endpoints, networks, identities, applications, and cloud audit trails. AI transforms investigation by automating data collection and correlation, reducing the time spent on manual log retrieval and normalization. When an incident is declared, an AI enabled workflow can automatically retrieve relevant telemetry for a defined

149

window before and after the triggering event. The system can then link events into chains, identify likely precursors, and flag anomalies that align with known attacker behaviors such as privilege escalation, persistence, and lateral movement. This produces an evolving narrative that responders can validate and refine, rather than a static set of alerts. The net effect is faster situational clarity under pressure.

AI is especially valuable in reconstructing attacker pathways and answering the core questions of Incident Response: how did entry occur, what did the adversary do, and what is the current risk. It can correlate identity anomalies with endpoint behavior, then map resulting access paths through network connections and cloud activity. It can identify suspicious process trees, rare command line usage patterns, and unusual parent child process relationships that often signal living off the land techniques. It can also connect user behavior analytics with asset criticality, helping responders understand whether the incident is limited to a workstation or has touched systems that support sensitive data or critical operations. This correlation supports both containment and eradication, because responders can prioritize actions based on confirmed pathways and likely objectives. The speed advantage is significant, but the precision advantage is often even more important.

Root cause analysis benefits from AI when the system is used to generate hypotheses that are grounded in evidence, rather than speculative attribution. For example, an AI workflow can propose that the incident began with credential compromise based on login anomalies, subsequent token misuse, and lateral movement patterns consistent with credential-based access. It can propose an initial phishing vector if email telemetry shows suspicious messages, user click behavior, and a subsequent payload execution timeline. It can propose exploitation of an unpatched service if application logs show abnormal requests and followed by process level anomalies on the host. These hypotheses should be framed as confidence rated conclusions supported by evidence, not absolute certainty. In practice, this speeds containment decisions and improves post incident learning because remediation targets the true enabling conditions.

Automated investigation must be designed to preserve forensic integrity and legal defensibility. AI systems should preserve raw evidence, record chain of custody for key artifacts, and log how conclusions were derived from data sources. They should minimize the risk of evidence contamination by separating investigation workflows from remediation workflows, particularly in high stakes incidents where litigation or regulatory reporting may follow. Where explainability is limited, the program should require supporting evidence links and independent validation before decisive actions are taken. This is one reason governance and process design matter as much as model capability. When the investigative workflow is evidence grounded and repeatable, AI becomes a reliable accelerant rather than a risky shortcut.

AI-Driven Response Recommendations and Orchestration

Once responders have clarity on scope and likely attacker objectives, the next constraint is the speed and quality of response decisions. AI can recommend containment actions by matching incident patterns to playbooks and using contextual signals to prioritize actions that reduce blast radius fastest. Recommendations are strongest when they are specific, such as isolating a particular endpoint, disabling a particular credential, blocking a specific destination, or limiting a service account to reduce privilege abuse. AI can also propose sequencing, such as disabling credentials first when identity compromise is suspected, then isolating endpoints to prevent further propagation. When recommendations are tied to evidence and confidence,

151

responders can move faster with less cognitive burden. This is especially valuable during multi vector incidents where parallel actions must be coordinated.

Orchestration improves when AI is integrated with SOAR capabilities that can coordinate actions across tools and teams. A well designed response pipeline can translate AI findings into structured tasks, create incident timelines, assign ownership, and track completion against service level objectives. AI can help route actions to the right team based on affected system ownership, business service mapping, and operational constraints. It can also identify dependencies that influence safe containment, such as whether isolating a server will disrupt a customer facing service. This enables a more precise containment strategy that reduces harm to operations while stopping the adversary. In effect, AI helps responders manage both the security problem and the business continuity problem in the same operational picture.

AI driven orchestration must be paired with strong decision authority and escalation design. Incident command must remain clearly defined, and AI should not become a shadow decision maker that bypasses established authority. Response recommendations should be reviewed in the context of risk appetite and approved action boundaries. Automation should be tiered so that low impact actions can be executed quickly while high impact actions require explicit approval. This is where governance moves from a policy statement to a practical control that prevents disruptive errors during crisis response. The objective is fast action with predictable control.

Response Automation and Human in the Loop Thresholds

Automation is the most powerful and risky lever in AI enabled Incident Response, because it can prevent rapid spread but it can also cause rapid disruption. The safest automation strategies are those that focus on reversible actions with well understood impact, such as blocking known malicious indicators, isolating a confirmed compromised endpoint, or forcing credential resets when high confidence compromise is detected. These actions can be executed quickly and rolled back if subsequent evidence changes the interpretation. High impact actions such as disabling critical services, quarantining core server segments, or blocking broad network paths should require human approval unless the organization has explicitly accepted the operational risk and tested rollback mechanisms. This tiered approach balances speed with safety. It also builds trust because responders see automation as a controlled capability rather than an unpredictable system.

Human in the loop design should be explicit, not informal, and it should be based on confidence thresholds and business impact categories. A practical model uses at least three tiers: automatic execution for high confidence and low impact actions, human approval required for moderate impact or moderate confidence actions, and incident commander approval required for high impact actions. Confidence should be defined through multiple signal corroboration rather than a single model score, and the program should define what counts as corroboration. For example, an identity compromise decision may require identity anomaly signals plus endpoint evidence or suspicious session behavior. This reduces the risk of acting on a single noisy signal. When thresholds are explicit, automation becomes a safe accelerator rather than a source of new incidents.

Automation also requires controls for reversibility, auditing, and safety stops. Every automated action should create a record of what was changed, why it was changed, and which evidence triggered the change. A rollback plan should be defined for common automated actions, and responders should be able to disable automation quickly if model behavior becomes untrusted or if the environment changes unexpectedly. This is a practical requirement, not a theoretical one, because large incidents often involve incomplete information and shifting interpretation over time. Automation should be treated like a high privilege actor with strict access control and monitoring. If the organization would not allow an unmonitored human to execute broad actions, it should not allow an unmonitored automated workflow to do so either.

Generating Incident Narratives, Timelines, and Reports with AI

Incident documentation is a critical part of response, not only for compliance but for decision quality during the incident. Leaders need a clear statement of what happened, what is impacted, what actions are underway, and what decisions are required. AI can generate incident narratives by synthesizing correlated evidence into a chronological account that updates as new information arrives. This reduces the reporting burden on responders and improves speed of communication to executives, legal counsel, and operational leaders. It also improves consistency across incidents, which makes trend analysis and readiness measurement more reliable. The key is that AI reporting should remain evidence grounded and avoid speculation.

Timelines are especially valuable because they clarify causality and reveal where delays occurred. AI can construct timelines that include attacker activity, defensive actions, and decision points, making it easier to understand which steps reduced risk and which steps created friction. A strong timeline also supports post incident review and audit readiness because it shows why key containment actions were taken and when. AI can tailor reporting to stakeholder needs by producing technical appendices for responders and executive summaries for leadership. The executive view should focus on scope, business impact, decisions, and current status, while the technical view should preserve indicators, artifacts, and investigative evidence. This reduces the risk that decision makers receive either too much detail or too little clarity.

Automated reporting must be governed to avoid two risks: accidental disclosure and inaccurate claims. Reports often include sensitive data such as identities, system names, indicators, and investigation artifacts, so access control and retention policies must be applied. AI systems must also avoid asserting attribution or intent without evidence because such statements can create legal and reputational exposure. The reporting workflow should clearly distinguish between observed facts, confidence rated inferences, and open questions. Human review should be required for external facing statements and for any language that implies wrongdoing by an employee, vendor, or partner. When these controls exist, AI reporting improves speed and consistency without undermining defensibility.

Post Incident Analysis and Learning with AI

The long-term value of Incident Response depends on whether the organization learns and measurably improves after each event. AI strengthens post incident learning by analyzing patterns across incidents, identifying systemic weaknesses, and validating whether remediation actually reduced risk. Instead of relying on anecdotal lessons, AI can quantify recurring entry vectors, repeated misconfigurations, and control gaps that appear across multiple events. It can also identify which alerts were early indicators of incidents and which signals were noisy, enabling tuning of detection pipelines. This shifts the organization from episodic improvement to continuous improvement. In effect, the IR program becomes a learning system rather than a collection of playbooks.

AI can also improve the quality of root cause remediation by connecting technical findings to control improvements. If incidents repeatedly involve credential abuse, AI can support prioritization of stronger identity controls and detection of session anomalies. If lateral movement repeatedly exploits weak segmentation, AI can help map exposure paths and recommend targeted segmentation improvements. If phishing repeatedly drives initial access, AI can support improvements in email security and user training targeted to observed lures. The benefit is that remediation becomes specific and evidence based rather than generic. This increases the likelihood that improvements will reduce recurrence rather than simply satisfying a checklist.

A mature learning loop also evaluates response performance, not only attacker behavior. AI can measure time to triage, time to containment, time to eradication, and time to recovery, then identify where delays occurred due to tooling gaps or process constraints. It can also analyze whether automation helped or harmed, and whether confidence thresholds were appropriate. These insights support governance decisions about where to expand automation and where to tighten controls. The post incident process should result in concrete changes to playbooks, controls, and training, and those changes should be tracked to confirm they produced measurable improvement. When AI is used in this way, it strengthens both security outcomes and operational discipline.

Ethical and Trustworthy AI in Incident Response

Trustworthy AI is essential in Incident Response because decisions are high stakes and errors can disrupt operations or unfairly implicate individuals. AI outputs should be explainable enough for responders to understand why a conclusion was reached, even when the underlying model is complex. Explainability does not require revealing every internal detail, but it does require evidence links, contributing factors, and confidence ratings. This helps responders validate AI findings rather than accept them as unquestionable. It also supports auditability and post incident review. In a crisis, trust is built through clarity and consistency, not through model sophistication alone.

Bias and fairness concerns also appear in Incident Response, especially where user behavior analytics and insider threat indicators are involved. AI can produce skewed risk scores if training data reflects historical patterns that correlate with organizational roles, locations, or work schedules in a misleading way. The program should monitor whether certain user groups or departments are disproportionately flagged and investigate whether the system is detecting real risk or simply modeling differences in work patterns. Where AI informs actions that affect user access or employment outcomes, governance must require human review and supporting evidence. Incident Response should focus on demonstrable behaviors and artifacts, not on inferred intent. This protects both the organization and the individuals involved.

Privacy and data protection are also central because IR workflows often ingest sensitive telemetry and personal information. AI systems should follow data minimization principles, limit access to sensitive fields, and apply retention policies aligned to regulatory and legal requirements. Sensitive outputs should be controlled, especially when AI generates narratives that could be shared widely. The organization should treat AI systems as privileged actors with strict access control, segmentation, and monitoring because the evidence pipeline itself becomes a high value target. Finally, the program should define manual fallback procedures so responders can continue to operate effectively if AI outputs are unavailable or untrusted. Trustworthy AI in IR is not only ethical, it is operationally necessary.

Key Risk Indicators and Operational Metrics for AI Enabled IR

Leadership needs measurable indicators that show whether AI is improving Incident Response without creating new risk. Core metrics include mean time to triage, mean time to containment, and mean time to recovery, measured by severity tier and tracked over time. AI specific metrics should include the accuracy of triage decisions, the false positive rate of AI driven escalation, and the false negative rate identified through post incident review. Automation metrics should include how often automated actions were executed, how often they were rolled back, and how often human overrides were used. These measures reveal whether automation is operating safely and whether confidence thresholds are correctly set. A program that cannot measure these outcomes cannot reliably govern them.

The organization should also track evidence quality and auditability. Metrics can include the percentage of incidents with complete timelines, the percentage of key conclusions supported by linked evidence, and the time required to produce executive summaries during high severity events. Drift monitoring is also important because changes in environment and attacker behavior can degrade model performance. If drift is not monitored, the program risks trusting outputs that are slowly becoming less accurate. Finally, training and readiness metrics matter because human capability remains decisive. The program should measure analyst adoption, ability to validate AI outputs, and performance during exercises that simulate AI failure or AI unavailability.

Implementation Guidance for Security Leaders

AI enabled Incident Response should be implemented as a controlled capability, not a tool purchase. The first step is defining the decision model: which decisions AI informs, which actions it may recommend, and which actions it may execute automatically. The second step is ensuring data integration and evidence pipelines are reliable, because AI cannot compensate for missing telemetry and inconsistent logging. The third step is designing playbooks that incorporate AI outputs in a structured way, including validation steps and escalation triggers. This is where process design determines whether AI reduces friction or creates confusion. A disciplined rollout focuses on narrow high value use cases first, then expands based on measured outcomes.

Automation should be introduced with special care and should prioritize reversible actions that reduce blast radius. Organizations should test automation in controlled simulations and tabletop exercises before allowing it to execute in production. They should also implement safety controls such as kill switches, rate limits, and rollback procedures, and they should monitor automation outcomes continuously. Governance should include an approval model, audit trails, and periodic review of confidence thresholds. Finally, the organization should invest in people, because AI shifts skill needs toward evidence reasoning, model validation, and orchestration rather than manual log searching. The goal is a human AI team that is faster, more consistent, and more resilient under pressure.

Board-Ready Brief

AI in Incident Response (IR): Speed, Control, and Organizational Resilience

Executive Context

Incident Response is the point at which cybersecurity risk becomes business risk. Regardless of how mature an organization's prevention and detection capabilities may be, the inevitability of incidents means that leadership outcomes are shaped by how quickly and effectively the organization responds under pressure. In modern hybrid environments, incidents escalate rapidly, often crossing identity systems, endpoints, cloud platforms, and third-party services within minutes. Artificial Intelligence has emerged as a critical capability for accelerating response, but its value is realized only when speed is balanced with governance, trust, and accountability.

This chapter focuses on how AI reshapes Incident Response from a reactive, labor-intensive process into a disciplined, intelligence-driven capability that reduces dwell time, limits operational disruption, and strengthens organizational resilience.

Why Incident Response Requires AI Augmentation

Traditional Incident Response models rely heavily on manual correlation, sequential investigation, and human interpretation of fragmented data. While these models remain conceptually sound, they struggle under modern conditions characterized by alert volume, attacker automation, and distributed infrastructure. The result is often delayed containment, incomplete scoping, and inconsistent response quality.

AI augments Incident Response by accelerating three critical dimensions:

- **Speed of understanding** through automated correlation and narrative generation

- **Quality of decisions** through contextual risk scoring and evidence-based recommendations

- **Consistency of execution** through governed automation and orchestration

These capabilities do not replace human leadership or judgment. Instead, they compress time to clarity, enabling responders and executives to act decisively with defensible evidence.

How AI Accelerates the Incident Response Lifecycle

AI enhances every phase of the Incident Response lifecycle, from detection through post-incident learning.

Early Detection and Triage

AI improves triage by correlating low-confidence signals into higher-confidence incident hypotheses. Identity anomalies, endpoint behaviors, and network indicators are analyzed together rather than in isolation. This reduces alert fatigue while increasing the likelihood that serious incidents are identified early.

Scoping and Impact Assessment

AI rapidly identifies affected systems, identities, and business services. This enables responders to focus containment on assets that matter most, rather than treating all compromised systems as equal risk.

Investigation and Root Cause Analysis

Automated evidence gathering and timeline construction reduce investigation time from hours to minutes. AI proposes root cause hypotheses grounded in observable data, supporting faster and more accurate remediation decisions.

Response Recommendation and Automation

AI recommends containment actions tailored to the incident context and can safely automate low-risk, reversible actions under predefined confidence thresholds. This is critical in fast-moving incidents where delay materially increases damage.

Post-Incident Learning

AI analyzes incidents collectively, identifying systemic

weaknesses, recurring attack paths, and control gaps. This transforms Incident Response into a continuous improvement engine rather than a series of isolated events.

Governance, Trust, and Risk Considerations

While AI materially improves Incident Response outcomes, it also introduces new governance responsibilities.

Key risks include:

- Over-automation without adequate confidence thresholds
- Opaque decision logic that undermines trust and auditability
- Bias in behavioral analytics affecting users or roles disproportionately
- Privacy exposure through expanded telemetry ingestion

A trustworthy AI-enabled IR program requires:

- Explicit human-in-the-loop decision boundaries
- Explainable outputs tied to evidence rather than opaque scores
- Auditable automation with rollback capability
- Privacy-aware data handling and retention controls

Boards should expect management to treat AI as a high-privilege actor, governed with the same rigor applied to financial systems or production change controls.

What the Board Should Expect to See

From a governance and oversight perspective, leadership should be able to demonstrate:

- Measurable reductions in time to triage, containment, and recovery

- Clear boundaries between automated actions and human approval
- Incident narratives that are timely, consistent, and evidence-based
- Metrics that track AI accuracy, automation effectiveness, and override frequency
- Ongoing testing of AI behavior through exercises and post-incident reviews

AI in Incident Response should be viewed as a resilience capability, not a technical enhancement. When governed effectively, it reduces uncertainty, improves decision quality under stress, and limits the operational and reputational impact of cyber incidents.

Conclusion: The AI-Augmented Incident Response

AI can materially strengthen Incident Response by compressing the time required to confirm incidents, scope impact, contain threats, and communicate clearly to leadership. It accelerates investigation through automated evidence gathering and correlation, improves decision quality through context aware recommendations, and enables safe automation when confidence and impact boundaries are explicitly governed. It also improves organizational learning by turning incidents into structured data that supports continuous improvement of controls, playbooks, and readiness. The strategic risk is that speed can amplify mistakes, which is why human in the loop thresholds, auditability, and rollback mechanisms are essential. Trustworthy AI in Incident Response requires explainable outputs, privacy disciplined evidence pipelines, fairness monitoring where user behavior is analyzed, and manual fallback procedures when AI outputs are unavailable or untrusted. When implemented as a governed capability, AI transforms Incident Response from a labor intensive scramble into a disciplined, faster, and more resilient business continuity function.

Chapter 7: Enhancing Security Metrics and Reporting with AI

Effective cybersecurity governance depends on measurement, yet many organizations measure what is easiest to count rather than what is most important to manage. Traditional security metrics often emphasize vulnerability totals, alert volume, patch velocity, and control implementation status, which can obscure actual exposure and mislead executive decision-making. These metrics are frequently lagging indicators, disconnected from business impact, and difficult to interpret without deep technical context. As a result, security reporting can become noisy, time-consuming, and strategically unhelpful, even when it is operationally detailed.

Artificial Intelligence enables a shift from activity reporting to decision-grade risk intelligence by correlating diverse telemetry, identifying meaningful patterns, and generating predictive indicators. This chapter examines how AI-driven metrics improve situational awareness for leadership by forecasting exploit likelihood, quantifying control effectiveness, and aligning security outcomes to business risk. It also addresses the governance requirements that accompany AI-based reporting, including transparency, confidence, auditability, and bias considerations. When implemented with discipline, AI-enhanced metrics strengthen both accountability and strategic clarity, allowing leadership to govern cybersecurity as an enterprise risk function rather than a collection of technical outputs.

The Limitations of Traditional Security Metrics

Traditional cybersecurity metrics were built for an earlier era when environments were smaller, threats were slower moving, and reporting cycles could lag operational reality without serious consequence. Many organizations still rely on familiar measures such as vulnerability totals, average patch time, alert volume, and counts of training completions, and these measures can be useful for operational hygiene. The problem is that they often fail to describe risk in a way that supports governance, prioritization, and investment decisions. When metrics emphasize activity rather than outcomes, leadership receives signals about how busy the security program is, but not whether the program is reducing the probability and impact of material cyber events. This creates a gap between operational reporting and executive decision-making, and it can leave boards and senior leaders governing security with incomplete or misleading evidence.

A primary weakness of many legacy metrics is that they function as lagging indicators, describing what happened after the fact instead of forecasting what is likely to happen next. A report that highlights the number of vulnerabilities discovered or the time to patch critical findings tells leadership that exposure existed and that remediation occurred, but it does not explain exploit likelihood, attacker intent, or the expected business impact if a vulnerability is exploited. A low vulnerability count can reflect a secure environment, yet it can also reflect weak scanning, incomplete asset inventory, or restricted visibility in cloud and third-party systems. A favorable average patch time may still hide unacceptable exposure if a handful of critical systems remain unpatched due to operational constraints. In practice, these metrics can look healthy while the organization remains vulnerable to the threats that matter most.

Traditional metrics also struggle to provide actionable context, especially for leaders who do not live in the daily flow of alerts and tickets. Alert counts are a common example. A high alert volume can indicate strong detection coverage, but it can also indicate poor tuning, noisy controls, and analyst fatigue. A low alert volume can signal stability, but it can also reflect blind spots, logging gaps, or quiet persistence by an adversary. Without correlation, confidence, and business translation, these numbers invite misinterpretation. Security teams can spend considerable time explaining what the data might mean, and even after those explanations, leadership may still lack clarity on whether risk is rising or falling.

A deeper limitation is the difficulty of tying technical security metrics to business outcomes. Cybersecurity exists to protect business services, sensitive information, operational continuity, and trust, yet many traditional measures operate in a technical vacuum. Executives need to understand which risks threaten revenue, availability, compliance posture, and brand, and they need decision options that align to risk appetite. Metrics that are not mapped to critical services and business impact are poor tools for enterprise governance. When security cannot express outcomes in business terms, it becomes harder to justify investments, harder to prioritize trade-offs, and easier for security to be treated as a cost center rather than a strategic enabler.

Another structural weakness is the overemphasis on input rather than outcomes. Counting patches applied, audits completed, firewall rules reviewed, or simulations run describes effort and throughput, but not necessarily effectiveness. A security program can produce impressive volume while failing to reduce real exposure, especially if teams are driven by checklists rather than risk-based prioritization. In these conditions, the organization can become highly efficient at completing tasks yet remain inefficient at reducing material risk. This is also where traditional metrics are most vulnerable to being "gamed," intentionally or unintentionally, by optimizing for appearances instead of outcomes.

Traditional measurements often fragment security reality into siloed categories, which obscures the interconnected nature of modern attacks. Endpoint metrics, email security metrics, cloud posture metrics, and identity metrics may each look acceptable in isolation, while a multi-stage attack path remains open across boundaries. Compliance-focused scoring can add further confusion by creating a false sense of security when "controls are implemented" but are not performing effectively, not instrumented properly, or not tuned for the threats most relevant to the organization. Modern adversaries exploit seams between controls, not the parts that are easiest to audit.

Finally, manual reporting cycles degrade the relevance of metrics in a fast-moving environment. If security reporting requires extensive manual collection, reconciliation, and formatting, leadership often seeing stale information that no longer matches the current threat landscape. This delay weakens decision velocity and increases the probability of governing with outdated assumptions. When the environment is dynamic and the adversary is adaptive, slow reporting is not merely inconvenient, it becomes a risk factor in itself.

Leveraging AI for Deeper Insight and Predictive Metrics

Artificial Intelligence changes the purpose and structure of security measurement by making it possible to transform massive volumes of heterogeneous telemetry into coherent, contextual, and forward-looking indicators. Instead of treating logs, alerts, and findings as separate streams, AI systems can ingest data across identity, endpoints, networks, cloud services, applications, and third-party integrations, then correlate it into unified narratives about exposure, adversary behavior, and defensive performance. This correlation capability is foundational because risk emerges from relationships across systems, not from isolated events. AI provides the scale and analytical continuity required to see those relationships in near real time, and it enables metrics that describe how risk is evolving rather than merely summarizing past activity.

A core advantage of AI-driven measurement is its ability to produce leading indicators rather than lagging counts. Predictive metrics are not about pretending the future is certain, but about improving the organization's ability to forecast where it is most likely to be harmed and to act before harm occurs. AI can combine vulnerability data with real-world exploitation signals, asset criticality, internet exposure, identity privilege levels, and observed attacker tactics to estimate exploit likelihood and expected impact. This enables dynamic risk scores that reflect current conditions rather than static averages. In practical terms, it shifts remediation from "patch the most severe findings" to "fix the findings most likely to be exploited against the assets that would produce the most damage."

AI also improves measurement by assessing control effectiveness as performance rather than presence. Traditional reporting might state that endpoint protection is deployed broadly, or that email security blocks a high percentage of known malicious messages, but these statements do not reveal how controls behave against novel threats, evasion tactics, and low-and-slow campaigns. AI can evaluate detection fidelity by observing outcomes over time, measuring false positives, false negatives, and time-to-detection for different attack patterns. It can also assess response performance by measuring how quickly an organization contains threats once detected, how consistently playbooks are executed, and where investigation stalls due to missing telemetry or excessive manual correlation. These metrics describe the health of the defensive system rather than the quantity of its outputs.

169

Another major contribution is the ability to quantify cyber resilience more directly. Resilience refers to the ability to withstand, adapt, and recover while maintaining critical operations, and it is not captured well by vulnerability totals or ticket throughput. AI-driven resilience indicators can incorporate incident response performance, recovery time objectives achieved in practice, coverage of critical services, and readiness of contingency controls. Resilience measurement becomes stronger when AI ties risk to business service dependency graphs, showing how technical failures cascade into operational impact. This aligns reporting with what leadership must govern: continuity, safety, compliance exposure, and financial risk.

AI-powered anomaly detection supports predictive measurement by establishing behavioral baselines for users, devices, applications, and services, then continuously evaluating deviations. These deviations can be translated into metrics that reflect risk posture in motion, such as the rate of high-confidence identity anomalies, the volume of suspicious privileged activity, or changes in data movement patterns across sensitive environments. Importantly, these are not merely "alert counts." They are weighted by context, confidence, and asset criticality, which is why they become useful as executive indicators. A small number of high-confidence anomalies involving privileged identities can matter more than thousands of low-confidence endpoint alerts, and AI makes that prioritization measurable.

AI can also generate decision-grade metrics from unstructured and semi-structured data, which traditional approaches largely ignore. Threat intelligence reports, vulnerability disclosures, dark web chatter, and vendor advisories contain signals about what adversaries are focusing on and what exploit chains are trending. Natural Language Processing can help extract relevant themes, map them to the organization's technology stack, and generate

170

indicators such as emerging threat relevance, exposure alignment, and time-to-mitigation for newly weaponized vulnerabilities. These are not speculative vanity metrics when handled responsibly. They are a way to reduce surprise by systematically tracking the threats most likely to intersect with the organization's real environment.

For executive governance, one of the most valuable capabilities is AI's ability to translate operational data into business-aligned risk statements. When AI-driven systems have access to asset inventories, service catalogs, and business criticality, they can describe security posture in terms leadership can act on. Metrics such as "critical service attack surface risk," "likelihood of material disruption to customer-facing systems," or "regulatory exposure forecast for high-sensitivity data environments" are more aligned to board decision-making than generic vulnerability counts. These indicators are most valuable when they are backed by traceable evidence and when they include confidence levels and key drivers, so leaders understand what is shaping the outcome.

AI also enables improved measurement of security investment effectiveness by correlating changes in control posture to changes in risk outcomes. Rather than reporting that a new tool was deployed, AI-enhanced measurement can track whether detection lead time improved, whether dwell time decreased, whether false positives dropped, whether containment speed increased, and whether critical risk concentrations moved in the desired direction. This supports stronger capital allocation decisions because leadership can see whether investments reduce exposure, improve responsiveness, or increase resilience. While cybersecurity ROI is never perfectly measured, AI improves the defensibility of the narrative by grounding it in observed operational outcomes rather than aspirational claims.

Improving the Quality and Transparency of Security Reporting

Security reporting is only as valuable as the decisions it enables. Many organizations have extensive reporting, yet leaders still lack clarity because the reports are too technical, too dense, or too disconnected from business outcomes. AI improves reporting quality by reducing noise, increasing context, and producing narratives that link security signals to risk decisions. The objective is not to produce more dashboards. The objective is to produce fewer, higher-quality indicators supported by traceable evidence that can withstand scrutiny.

AI enhances report clarity by summarizing complex multi-source data into concise insights. Instead of listing alerts and vulnerabilities, AI-driven reports can identify what changed, why it changed, and what it means for risk posture. This is where automated correlation becomes especially important. AI can explain when multiple technical symptoms represent a single underlying issue, such as identity sprawl driving excessive privilege risk, or misconfigured cloud storage driving data exposure. Reducing duplication and fragmentation improves trust in the reporting because leadership sees coherent explanations rather than conflicting counts.

Predictive reporting is another quality improvement because it shifts leadership conversations from hindsight to preparedness. A board does not only need to know what happened last quarter. It needs to know what management believes is most likely to happen next, and what actions are underway to reduce that probability. AI can provide evidence-based risk forecasts by combining internal posture data with external exploitation signals and adversary behavior trends. A credible report does not present these forecasts as certainty. It presents them as risk ranges with confidence levels, key drivers, and recommended mitigation options. This is a governance-friendly way to discuss uncertainty without losing decision utility.

Transparency is essential because AI reporting can otherwise look like a black box. High-quality AI-driven reporting includes explainability elements: why a risk score changed, what data sources contributed, and what evidence supports the conclusion. When a report indicates rising identity risk in a region, leadership should be able to see whether the driver is increased privileged access, abnormal authentication patterns, new third-party integrations, or decreased logging fidelity. When reporting makes causal drivers visible, it becomes a tool for accountability, not merely a status update. It also reduces the risk of over-trusting AI outputs because stakeholders can see how conclusions were formed.

Business alignment is improved when reporting is structured around critical services and material risk. AI can support this by mapping security telemetry to service ownership, business processes, and data classifications. Reporting then becomes more actionable because leaders can allocate resources to places where risk is concentrated rather than responding to whichever dashboard is loudest. Boards can also see whether the security program is protecting the most important outcomes, such as uptime for customer platforms, integrity of financial systems, safety of operational technology, and compliance posture in regulated environments.

AI-driven reporting also supports role-based personalization without fragmenting truth. Executives need strategic indicators and decision options. Business unit leaders need targeted exposure views tied to their operations. Security and technology leaders need diagnostic detail to drive action. AI can tailor the presentation layer while keeping a consistent underlying evidence base, which improves trust across stakeholders. This reduces the common problem where different audiences receive different numbers because each report was generated manually or using different filters.

Visualization improves report comprehension, but AI's real contribution is intelligent visualization rather than decorative charts. AI can generate heat maps of risk concentration, trend lines showing posture movement, and attack-path diagrams showing how exposures connect across identity and infrastructure. These visuals become decision tools when they are tied to risk drivers and remediation options. Leaders should be able to see what matters, why it matters, and what it will take to improve it. The best reporting does not merely describe risk; it enables informed trade-offs.

Automation also improves timeliness. When AI reduces manual collection and synthesis, reporting can be delivered more frequently without burdening teams. This supports faster decision cycles, which is increasingly important in environments where threat conditions change weekly rather than quarterly. Timeliness also supports continuous improvement because leaders can observe whether changes in policy, controls, or staffing are producing measurable impact. Reporting becomes part of operational governance rather than a periodic compliance ritual.

Quantifying the ROI of AI in Cybersecurity Metrics and Reporting

Quantifying the value of AI in security measurement is less about claiming that AI "prevents breaches" and more about demonstrating how AI improves decision quality, reduces operational waste, and accelerates risk reduction. The most credible ROI narratives are grounded in specific, observable changes to security performance and in avoided costs that can be estimated with defensible assumptions. AI strengthens this conversation because it can measure improvements that were previously difficult to capture, such as reductions in investigative effort, increases in detection precision, and better prioritization of remediation work.

Operational efficiency is often the earliest measurable ROI. When AI reduces false positives and improves alert prioritization, analysts spend less time on noise and more time on high-value work. This efficiency can be quantified through reductions in time spent per alert, reductions in mean time to triage, and decreases in backlog growth. Reporting efficiency also improves when AI automates data aggregation, reconciliation, and narrative generation. If monthly reporting previously required extensive manual effort, AI can reduce preparation time while increasing consistency, allowing security leadership to spend more time on risk decisions rather than report assembly.

Risk reduction ROI can be articulated through improved exposure management. When AI prioritizes vulnerabilities based on exploitability and asset criticality, organizations can reduce the most dangerous exposure faster. ROI can be measured by tracking reductions in critical exploitable vulnerabilities, reductions in exposure windows, and decreases in attack-path completeness for critical services. When these indicators improve, the organization can credibly argue that the likelihood of a material event has decreased. This is not a guarantee, but it is defensible risk management progress, and boards govern risk, not certainty.

Incident cost reduction is another ROI pathway, even for metrics and reporting improvements. Better measurement improves preparedness and decision timing. If AI-driven reporting identifies rising risk in a specific control domain early, leadership can invest before a disruption occurs. AI can also reduce dwell time by improving detection fidelity and helping prioritize response actions. Reduced dwell time correlates with reduced impact, including fewer systems affected, less data accessed, and shorter recovery cycles.

Organizations can estimate avoided costs using incident modeling, including downtime, response labor, external support, regulatory exposure, and reputational harm, while clearly stating assumptions.

Compliance and audit readiness often provide concrete ROI evidence. If AI-driven reporting improves evidence traceability and continuous control monitoring, audit preparation can be faster and less disruptive. This reduces internal labor cost and external consulting dependence, and it lowers the risk of findings caused by incomplete documentation. AI can also enable early warnings of compliance drift by identifying misconfigurations and access anomalies that tend to produce violations. Preventing compliance failures has measurable value in avoiding fines, operational restrictions, and costly remediation programs.

The most sustainable ROI story connects AI-driven metrics to better capital allocation. When security leaders can demonstrate, with evidence, which investments move risk indicators in the desired direction, budgets become easier to defend and easier to optimize. This elevates security from reactive spending to managed investment. AI does not create this clarity automatically, but it makes it feasible if the organization designs reporting around risk drivers, performance measures, and governance needs.

Communicating AI-Driven Security Insights to the Board

Boards govern cybersecurity as enterprise risk. They do not need to understand how models work at a technical level, but they must be able to understand what the organization believes about risk, how confident it is, what actions are underway, and what decisions require board support. AI-driven insights are valuable only when they are translated into a governance-friendly narrative. This requires a disciplined reporting structure that minimizes jargon, emphasizes material outcomes, and clearly states the decision implications.

Effective board communication begins by framing AI-driven metrics in terms of business impact and risk posture, not tool performance. Rather than reporting that AI detected a certain number of anomalies, leadership should report how AI changed detection lead time, reduced alert noise, improved prioritization of critical exposures, and increased confidence that the organization is seeing the threats that matter. The conversation should focus on the health of critical services, the protection of sensitive data, and the organization's readiness to respond to disruptive events. Boards respond to risk narratives supported by evidence, not to technical detail.

A board-facing report should be anchored in a small number of enterprise indicators that are stable over time. These might include an overall cyber risk trend, risk concentration by critical service, resilience indicators such as recovery readiness, and exposure indicators such as attack-path completeness. AI should be described as the mechanism that improves the quality of these indicators by enhancing correlation, increasing context, and enabling prediction. The board should also see what changed since the last reporting cycle, why it changed, and what management is doing about it.

Visualizations should be used to increase comprehension, not to impress. Heat maps of risk concentration, trend lines of posture movement, and high-level attack-path diagrams can be effective when they remain tied to business services and remediation actions. The board should be able to interpret the visual in seconds and ask informed questions. If the visualization requires lengthy explanation, it belongs in an operational report rather than a board packet.

A responsible approach also requires communicating limitations. AI can produce false positives and false negatives, and it can drift as environments change. The board should be told how the organization manages these risks through model monitoring, evidence traceability, human review thresholds, and auditability. This transparency increases trust and prevents AI from becoming a credibility liability. Boards are increasingly attentive to responsible AI governance, and security leaders should be prepared to explain how privacy, bias, and accountability are managed within AI-based analytics, especially when identity and behavior data are used.

The most important outcome of board communication is decision alignment. Each report should end with clear decision points, such as investments needed to address concentrated risk, policy choices that affect risk tolerance, and resource constraints that impede remediation. When AI-driven reporting is well-designed, it reduces ambiguity and accelerates governance decisions. It shifts board conversations away from confusion about what the numbers mean and toward informed trade-offs about what the organization will do next.

Board-Ready Brief

Enhancing Security Metrics and Reporting with AI: From Activity Counts to Decision-Grade Risk Insight

Executive Context

Security measurement is one of the most common failure points in cybersecurity governance, not because metrics are absent, but because many metrics do not translate into decision-grade insight. Traditional measures such as vulnerability counts, patch time averages, alert volume, and training completion rates often describe activity rather than risk reduction. These indicators can be directionally useful, yet they are frequently lagging, easily misinterpreted, and difficult to connect to business impact. Boards are then forced to govern cybersecurity using reports that are technically dense, operationally noisy, and strategically ambiguous.

This chapter explains how Artificial Intelligence modernizes security measurement by shifting organizations from retrospective activity reporting to predictive, contextual, and business-aligned risk intelligence.

Why Traditional Metrics Fail at Board-Level Oversight

Many legacy cybersecurity metrics were designed for operations teams, not for enterprise risk governance. They often measure inputs, such as "how much work occurred," rather than outcomes, such as "how much risk was reduced." They are also prone to false confidence because a low vulnerability count may reflect weak scanning, and a low alert count may reflect blind spots rather than safety. When metrics lack context, executives are forced to interpret raw numbers without knowing whether they indicate progress, drift, or exposure.

From a governance standpoint, the central issue is that traditional metrics rarely answer the questions boards need to govern; what is the current risk posture, where is risk concentrated, what is changing, and what decisions are required.

What AI Changes in Security Metrics and Reporting

AI improves security measurement by turning large volumes of fragmented telemetry into interpretable, correlated, and predictive insight. Instead of reporting what happened last month, AI-driven metrics can forecast what is likely to happen next based on exploit activity, asset criticality, identity exposure, control effectiveness, and observed adversary behavior. This produces leading indicators such as prioritized risk trends, predictive compliance deviation signals, and confidence-weighted attack surface exposure. It also enables better measurement of security control performance, including how well controls detect novel attacks and how quickly teams can contain emerging threats.

AI makes reporting more actionable by converting operational complexity into concise narratives tied to business impact and decision options.

What the Board Should Expect to See

A mature AI-enabled security reporting capability should deliver fewer metrics, but each metric should be higher quality and more defensible. Boards should expect security reporting to include risk concentration views, trend direction, and confidence levels, rather than static counts. Reporting should connect cyber risk to business outcomes such as downtime likelihood, regulatory exposure, operational continuity, and financial loss scenarios. Security leaders should also be able to explain the governance model behind AI-driven metrics, including how models are monitored for drift, how false positives are handled, and how decisions remain auditable.

AI does not eliminate uncertainty, but it reduces ambiguity and strengthens the organization's ability to govern security with clarity.

Key Governance and Trust Considerations

AI-driven metrics introduce new responsibilities because model outputs can be misunderstood, over-trusted, or treated as objective truth. Boards should require transparency in how risk scores are calculated, what data sources are included, and what limitations exist. Confidence thresholds, auditability, and human review processes must be explicit, especially when AI-derived reporting influences investment decisions or performance assessments. Privacy, bias, and role-based impact analysis are also essential when AI uses identity and behavior analytics.

The objective is trustworthy measurement: fast insight paired with defensible reasoning and accountable oversight.

Board Actions and Oversight Questions

Boards can strengthen governance by asking for a short, consistent set of oversight questions for each reporting cycle. Are risk indicators trending in the right direction for the most critical business services, and what is driving the trend? Which top risks have the highest likelihood and the highest impact, and what mitigation options exist? What evidence supports the risk scores, and how is model quality monitored over time? Where does AI improve speed and clarity, and where is human judgment still required? What decisions does management recommend based on the current reporting, and what resources are needed to execute them?

This chapter positions metrics and reporting as a strategic control system, not a compliance artifact.

Conclusion: Security Metrics and Reporting

The limitations of traditional security metrics are not merely a reporting problem, they are a governance problem. When leaders are presented with activity counts instead of risk insight, they cannot reliably prioritize, invest, or oversee cybersecurity as an enterprise function. Artificial Intelligence enables a more mature measurement model by correlating telemetry across domains, producing predictive indicators, and translating operational complexity into decision-grade reporting aligned to business impact. The value of AI-driven metrics is realized when reporting becomes more contextual, more timely, more transparent, and more actionable, allowing leadership to govern cyber risk with greater clarity and confidence.

At the same time, AI-driven reporting introduces new responsibilities. Model outputs must be explainable, auditable, monitored for drift, and governed to protect privacy and prevent inappropriate conclusions. When implemented with discipline, AI enhances both accountability and strategic alignment, turning metrics into a control system for risk management rather than a catalog of operational activity. The future of security measurement belongs to organizations that treat reporting as a strategic capability and build AI-enabled metrics that are defensible, business-relevant, and designed to support governance decisions at speed.

Chapter 8: Building the Human-AI Teaming Model

The accelerating pace and complexity of cyber threats have fundamentally changed what security teams must accomplish in a limited window of time. Security operations now involve continuous monitoring across endpoints, networks, cloud services, identities, applications, and third-party ecosystems, all producing high-volume telemetry with uneven signal quality. Human expertise remains essential, but humans cannot scale to machine-speed attack cycles or manually correlate millions of weak signals across dozens of systems. Artificial intelligence can scale and correlate, but it cannot fully understand intent, business context, ethical boundaries, or operational tradeoffs. The result is a persistent tension: without AI, defenders fall behind the pace of threats; without humans, AI-driven decisions can create new forms of risk.

Human–AI teaming resolves this tension by treating AI as an amplifier of expertise rather than a replacement for judgment. In a mature teaming model, AI performs rapid triage, correlation, summarization, and hypothesis generation. Human analysts validate, interpret, prioritize, and authorize response actions, particularly when decisions affect safety, business continuity, user access, or regulatory posture. This chapter provides a practical, operational blueprint for building a human–AI security team that is trusted, explainable, and auditable. It focuses on the mechanisms that make teaming work in reality: workflow design, defined decision tiers, accountability structures, governance controls, cultural trust, and measurable outcomes.

This is not a "tools chapter." It is a chapter about operating discipline. The goal is to help organizations avoid the two most common failure modes: deploying AI in ways that produce noise and erode trust or deploying automation in ways that undermine accountability. When human–AI teaming is designed well, security teams become faster, more consistent, and more resilient without surrendering decision authority.

The Synergy of Human Expertise and AI Capabilities

The effectiveness of human–AI teaming begins with a clear understanding that humans and artificial intelligence excel at fundamentally different forms of cognition. AI systems are optimized for scale, repetition, and statistical pattern recognition. They can process enormous volumes of data without fatigue, correlate activity across disparate systems, and surface weak signals that would remain invisible to manual analysis. Humans, by contrast, are optimized for contextual reasoning. They interpret meaning, infer intent, evaluate consequences, and navigate ambiguity in ways that machines cannot reliably replicate. Security operations improve not when these capabilities overlap, but when they are intentionally aligned.

In practice, AI's most valuable contribution is not decision-making, but sensemaking. It transforms raw telemetry into structured representations of activity, assembling timelines, clustering related events, and highlighting deviations from established norms. This transformation reduces cognitive overload and allows human analysts to engage with security problems at a higher level of abstraction. Instead of reacting to individual alerts, analysts evaluate narratives: what appears to be happening, how activity is evolving, and which hypotheses best explain observed behavior. AI thus changes the nature of human work from signal detection to judgment and prioritization.

Human judgment remains indispensable because security decisions are never purely technical. Analysts must consider business context, operational dependencies, user roles, regulatory obligations, and potential second-order effects. Anomalous behavior may reflect malicious intent, but it may also reflect legitimate change driven by business events such

184

as mergers, reorganizations, or emergency response activities. AI can highlight deviation, but it cannot reliably determine meaning. Humans apply situational awareness, institutional knowledge, and ethical reasoning to distinguish threat from noise and response from overreaction.

The synergy deepens over time as humans and AI learn from each other. Human analysts correct misclassifications, identify novel attack techniques, and recognize emerging adversary behaviors that fall outside historical training data. When this feedback is systematically incorporated, AI models evolve, detection improves, and analyst trust increases. This iterative learning process is critical. Without it, AI remains static while threats evolve, and human confidence erodes. With it, the partnership becomes adaptive, reinforcing both machine capability and human effectiveness.

Designing Workflows for Effective Human-AI Collaboration

Human–AI teaming succeeds or fails at the workflow layer. Even highly accurate AI systems will degrade operational performance if their outputs arrive without context, explanation, or clear paths to action. Effective collaboration requires workflows that explicitly define how AI and humans interact across the security lifecycle, from detection through resolution and learning.

In a coherent teaming model, AI functions as the first stage of analytical compression. It ingests raw data, removes duplication, enriches signals with contextual attributes, and assembles related activity into coherent investigative artifacts. These artifacts must be stable in structure and language so analysts can rapidly interpret them without reorienting cognitively. When AI outputs are inconsistent or overly technical, analysts expend effort translating information rather than applying judgment, negating the intended benefit of automation.

Human involvement intensifies as decisions carry greater consequence. AI may recommend investigative focus areas or propose response actions based on observed patterns and learned correlations. Humans evaluate these recommendations through the lens of operational risk, business impact, and organizational tolerance for disruption. The distinction between recommendation and authority is critical. AI informs; humans decide. This boundary preserves accountability while still capturing the speed and analytical depth AI provides.

Workflow coherence also depends on disciplined transitions between roles. As incidents escalate, information must flow seamlessly from analysts to incident commanders, legal advisors, and executives. AI can reduce friction by generating consistent, narrative summaries that capture what is known, what remains uncertain, and what actions are proposed. When handoffs are structured and intelligible, decision-makers receive clarity rather than raw data, enabling timely and defensible responses.

Explainability underpins the entire workflow. Analysts must understand why AI flagged an event, which factors contributed most strongly, and how confident the system is in its assessment. Explainability is not an optional enhancement; it is the mechanism that enables trust, oversight, and improvement. Without it, AI becomes either blindly followed or reflexively ignored. With it, AI becomes a credible analytical partner whose outputs can be scrutinized, refined, and relied upon appropriately.

Maintaining Human Accountability in AI-Assisted Decisions

As AI becomes embedded in security operations, preserving human accountability becomes a central governance obligation. Accountability does not persist simply because humans are involved. It persists only when authority, responsibility, and decision ownership are explicitly defined and reinforced through process.

In a mature teaming model, every class of AI-informed decision has a clearly identified human owner. Analysts validate detections, but incident commanders authorize containment strategies. Identity and access owners approve changes affecting personnel. Governance and legal stakeholders define automation boundaries and escalation thresholds. This clarity prevents the diffusion of responsibility that often accompanies automated systems, where recommendations are executed without clear ownership of outcomes.

High-impact decisions require traceability. When AI recommendations influence actions that affect availability, access, or regulatory exposure, the rationale for those decisions must be documented. This documentation is not merely administrative; it is essential for auditability, post-incident learning, and organizational accountability. Recording how evidence was interpreted and why actions were taken ensures that security decisions remain defensible even under external scrutiny.

Human override must be institutionalized rather than discouraged. Analysts must be trained and culturally supported to challenge AI outputs when context, ethics, or operational judgment indicate risk. Overrides should be examined systematically, not treated as anomalies. Patterns in overrides often reveal gaps in training data, missing contextual inputs, or overly aggressive automation thresholds. Studying these patterns strengthens both human judgment and AI performance.

Accountability also requires active management of automation expansion. Over time, confidence in AI systems can lead to gradual increases in autonomous action without corresponding reassessment of risk. Preventing this automation drift requires periodic review, rollback testing, and governance oversight to ensure that autonomy remains aligned with organizational risk tolerance.

Fostering a Culture of Trust and Collaboration

Trust is the invisible infrastructure that determines whether human–AI teaming succeeds or fails. Analysts must trust that AI outputs are reliable, interpretable, and aligned with organizational values. AI systems, in turn, must be designed to respect human judgment rather than obscure or bypass it.

Predictability is a foundational element of trust. Analysts develop confidence in systems that behave consistently in how they score risk, present evidence, and escalate severity. Volatile or opaque behavior undermines trust even when detection accuracy is technically high. Consistency allows analysts to calibrate their reliance appropriately.

Feedback is equally critical. Analysts disengage when corrections disappear into a void. When feedback loops are visible and produce tangible improvements, trust deepens. Ethical considerations further shape trust, particularly in areas such as insider risk and behavioral analytics. Monitoring systems must adhere to principles of proportionality, purpose limitation, and fairness. Human review must precede punitive or personnel-impacting actions, reinforcing legitimacy and ethical governance.

Leadership plays a decisive role in shaping trust. When AI is framed as a workforce reduction mechanism, collaboration gives way to resistance. When it is framed as a means of reducing noise, elevating expertise, and improving resilience, adoption accelerates. Leaders reinforce this framing through training investment, staffing decisions, and recognition of successful human–AI collaboration.

Measuring the Effectiveness of Human-AI Teams

The effectiveness of human–AI teaming cannot be assessed through isolated tool metrics. What matters is the performance of the combined system. Operational indicators such as detection speed, response time, and alert volume reduction reveal whether AI is meaningfully reducing cognitive load. Quality indicators such as false positive reduction, false negative discovery, and override accuracy reflect whether human judgment and machine intelligence are reinforcing each other.

Workforce health metrics provide additional insight. Effective teaming shifts analyst effort from reactive triage toward proactive threat hunting and strategic analysis. Burnout indicators decline, confidence increases, and retention stabilizes. Governance metrics confirm that accountability has not been sacrificed for speed, demonstrating that automation remains bounded by oversight.

Taken together, these measures indicate whether human–AI teaming is producing resilience rather than the illusion of progress. The goal is not simply faster action, but consistently better decisions under pressure.

Board Ready Brief

Building the Human–AI Teaming Model: Governance, Trust, and Operational Advantage

Executive Context

Cybersecurity effectiveness is no longer determined solely by tools or headcount, but by how well humans and intelligent systems operate together under pressure. Artificial Intelligence has introduced unprecedented speed, scale, and analytical power into security operations, yet these capabilities only produce value when paired with disciplined human judgment, accountability, and contextual reasoning. Organizations that treat AI as a replacement for expertise often introduce new risks, including automation errors, ethical failures, and loss of operational trust. Organizations that treat AI as an augmentation to human decision-making gain resilience, speed, and strategic clarity.

This chapter frames human–AI teaming as a governance and operating model, not a technology deployment. It explains how leadership can ensure AI strengthens security outcomes while preserving accountability, trust, and ethical control.

Why Human–AI Teaming Is Now a Leadership Issue

The scale and velocity of modern cyber threats exceed what human teams can process unaided. Telemetry volumes, attack automation, and adversary sophistication demand machine-speed analysis. At the same time, AI systems lack contextual understanding, ethical judgment, and accountability. When AI operates without structured human oversight, it can amplify errors at machine speed.

Human–AI teaming resolves this tension by assigning AI to tasks where speed, pattern recognition, and scale dominate, while reserving judgment, prioritization, ethics, and final authority for humans. This is not a compromise; it is a design principle that produces stronger outcomes than either humans or machines operating independently.

What the Board Should Expect from a Mature Human–AI Model

Boards should expect clarity, not mystique. A mature human–AI security model demonstrates where AI operates autonomously, where it advises, and where humans retain decision authority. Leaders should be able to explain how AI outputs are validated, how bias and drift are monitored, and how accountability is preserved when AI recommendations influence action. The model should improve detection speed and consistency while reducing analyst fatigue and error rates.

Boards should also expect workforce enablement, not displacement. Effective teaming elevates human roles toward higher-value analysis, strategy, and response coordination. The outcome is not fewer people, but more capable people operating with better information and stronger controls.

Governance, Ethics, and Trust

Human–AI teaming introduces governance obligations. AI recommendations must be explainable, auditable, and challengeable. Decisions that affect people, business operations, or regulatory posture must remain human-owned. Ethical safeguards must address bias, privacy, and proportionality, especially in identity and behavior analytics.

Trust is not automatic. It is built through transparency, feedback loops, and leadership reinforcement that AI is a decision-support system, not a decision-maker. When trust is present, teams act faster and with greater confidence. When trust erodes, AI becomes noise rather than leverage.

Board Oversight Questions

- Where does AI act autonomously, and where is human approval required?
- How are AI recommendations explained and audited?
- How is bias monitored, especially in identity and insider-risk analytics?
- Are humans trained to challenge AI when judgment or ethics demand it?
- How does human–AI teaming improve resilience, not just efficiency?

Human–AI teaming is ultimately a leadership design choice. Done well, it creates a security organization that is faster, more consistent, and more resilient without surrendering control or accountability.

Conclusion: Human-AI Teaming Model

Human–AI teaming is not a technology trend; it is an operating model for modern cybersecurity. AI provides speed, scale, and analytical depth, while humans provide judgment, ethics, and accountability. When these strengths are deliberately aligned, organizations gain resilience rather than fragility, confidence rather than automation risk.

Effective human–AI teams are built through intentional workflow design, clear accountability, explainable systems, and a culture of trust. Leadership plays a decisive role by framing AI as an amplifier of expertise rather than a replacement for responsibility. When governance is clear and collaboration is strong, human–AI teaming transforms security from a reactive function into a disciplined, adaptive defense capability.

The organizations that succeed will not be those with the most automation, but those that design the most effective partnerships between intelligent systems and accountable human judgment.

Chapter 9: Compliance at Scale with AI Assistance

Compliance has always been a discipline of proof. It is not enough to claim that controls exist, or even that they are well-designed. Organizations must demonstrate, with defensible evidence, that controls are implemented consistently, operating effectively, and producing outcomes aligned to regulatory expectations and internal policy commitments. In earlier eras, this proof could be assembled through periodic audits, interviews, sampling, and documentation reviews. That approach was never elegant, but it was workable when systems were slower-changing, data flows were more contained, and control environments were comparatively stable.

Artificial intelligence disrupts each of those assumptions. AI systems evolve, operate at scale, and increasingly influence decisions that affect individuals, customers, employees, and markets. The compliance problem is no longer limited to whether a system is configured correctly at a point in time. It extends to whether decisions can be explained, whether behavior remains bounded as models drift, whether data practices remain lawful as training sets expand, and whether accountability remains intact as automation becomes more consequential. The compliance function is therefore being forced into a new identity. It must move from periodic assurance to continuous governance. It must move from document collection to system-level observability. It must move from static control mapping to dynamic validation across fast-changing environments.

This chapter builds a coherent model for achieving compliance at scale with AI assistance, while avoiding the common failure mode of "automation without defensibility." AI can accelerate evidence capture, control validation, reporting, and monitoring, but only if its use is designed around auditability, transparency, and human accountability.

Compliance at scale is not achieved by generating more dashboards or longer reports. It is achieved by building a verifiable chain from requirement to control, from control to telemetry, from telemetry to evidence, and from evidence to narrative assurance that stands up under scrutiny.

The Compliance Challenge in the AI Era

The defining compliance tension of the AI era is that regulators and auditors expect systems to be explainable and stable, while modern AI is often probabilistic and adaptive. Traditional compliance programs are structured around artifacts: policies, procedures, system configurations, and sampled evidence. Those artifacts work reasonably well when the system's behavior is the predictable consequence of static code and controlled change processes. AI complicates this because behavior may emerge from learned patterns rather than explicit rules, and may shift over time through retraining, data drift, model updates, or changes in the underlying operational environment.

This shift produces several interlocking pressure points. The most visible is interpretability. Many AI models, particularly those built on deep learning or complex ensemble approaches, do not naturally produce the kind of decision trace that compliance regimes often assume. When an AI flags a transaction as suspicious, routes a customer for heightened verification, or influences a security action such as isolating a device, compliance stakeholders may need to explain why that output occurred. In privacy and fairness contexts, the need for explainability becomes even sharper because the impact of AI is directly human-facing. When a system meaningfully affects a person, the compliance obligation extends beyond "the model is accurate." It reaches into "the decision was lawful, justified, proportionate, and not discriminatory," and it must be documented in a way that can be reviewed and defended.

194

A second pressure point is continuous change. AI models are not merely deployed; they are managed across lifecycles. They may be retrained, fine-tuned, updated, or shifted into new contexts. Even without explicit retraining, models can experience drift as the data they encounter changes. From a compliance perspective, drift is not only a performance issue. It is a governance issue. A model that was compliant during an assessment may become noncompliant as its outputs change, as its false positives increase, as its bias profile shifts, or as it begins to rely on new data features that were not part of the approved design. The organization can no longer rely on "audit readiness" as a seasonal activity. Compliance must become a continuously sustained condition.

A third pressure point is data. AI systems expand the compliance surface area because they intensify data dependence. Training data, inference data, model outputs, feature stores, embeddings, and logs become part of the compliance environment. Privacy laws and security frameworks generally assume that organizations can account for where data comes from, how it is used, who can access it, how long it is retained, and how it is protected. AI makes each of those questions harder because datasets are larger, provenance is more complex, and inference outputs can reveal sensitive information through correlation. Even anonymization strategies can fail when AI is capable of linking data across sources. Compliance therefore requires data governance that is explicitly AI-aware, including controls around minimization, lawful basis, retention, access restrictions, and safeguards against reidentification.

A fourth pressure point is accountability. When AI is introduced into business processes, responsibility can become diluted across data science teams, engineering teams, business owners, vendors, and operational stakeholders. Compliance failure thrives in ambiguity. If no single role owns the system's compliance posture end-to-end, evidence becomes incomplete, remediation becomes slow, and audit narratives become fragile. Effective compliance at scale requires an explicit accountability architecture: roles, approvals, escalation pathways, and decision rights that prevent AI systems from becoming "everyone's tool and no one's responsibility."

A final pressure point is jurisdiction and fragmentation. Many organizations operate across multiple regulatory regimes and standards simultaneously. AI deployments often span cloud platforms, third-party services, and cross-border data flows. Compliance cannot be assumed to generalize. A model that is acceptable in one jurisdiction may violate requirements in another due to differing consent standards, data localization rules, or interpretations of automated decision-making. This does not mean AI is incompatible with global operations. It means compliance must be engineered to be adaptable, with controls that can be parameterized by region, data classification, and use case rather than assumed to be uniform.

Taken together, these pressures explain why AI-era compliance is not a straightforward extension of traditional GRC practices. It is a redesign. It is a shift from policy-to-control mapping as a documentation exercise toward control-to-behavior validation as a living system.

AI for Evidence Mapping and Control Validation

Compliance breaks down at scale not because organizations lack policies, but because they cannot sustain evidence of integrity across complex environments. Evidence mapping is the work of connecting compliance requirements to demonstrable proof: logs, configurations, approvals, access records, monitoring reports, incident tickets, training attestations, and system outputs. In large organizations, this becomes a constant struggle because evidence is distributed across tools, teams, and time. Manual evidence collection tends to produce three predictable outcomes: it is slow, it is inconsistent, and it is incomplete. The result is a compliance posture that appears strong on paper but weak under audit scrutiny.

AI can improve this condition if it is used as an evidence correlation engine rather than merely a report generator. The central value proposition is that AI can ingest large volumes of enterprise telemetry and documentation, classify it according to control requirements, and continuously update an evidence inventory that remains aligned to the current environment. This matters because many compliance controls are not satisfied by a single artifact. They require proof of implementation, proof of operation, proof of oversight, and proof of correction when failures occur. AI can link these layers, producing evidence chains instead of evidence fragments.

For example, a control requiring privileged access oversight is not satisfied only by the existence of an access policy. It requires identity system configuration evidence, access event logs, review records, exception approvals, and monitoring outputs that show the process is functioning. AI-enabled evidence mapping can correlate privileged role assignments with actual usage events, detect anomalies such as privileged logins occurring outside expected patterns, and surface whether mandated reviews occurred on schedule. The compliance team's work changes from hunting for artifacts to evaluating the system's demonstrated behavior and strengthening weak points.

Control validation becomes far more powerful when it shifts from periodic sampling to continuous verification. Traditional audits often rely on sample-based evidence, which can be necessary but also misleading. Systems can pass an audit while failing intermittently, or while drifting into noncompliance shortly after the assessment window. AI can help establish a continuous validation loop by monitoring control signals that indicate whether a control is functioning. In configuration management, this includes baseline drift detection. In patch management, this includes monitoring time-to-remediate against policy thresholds. In access controls, this includes usage patterns, unauthorized attempts, and deviations from least-privilege expectations. In data protection, this includes detection of unencrypted storage, unexpected data movement, and retention violations.

The key to making AI-based validation defensible is that it must be explainable in audit terms. AI should not merely conclude that a control is "effective." It must show what it evaluated, what evidence supports the conclusion, and where uncertainty exists. Compliance requires proof, not confidence scores. Therefore, AI-driven validation is strongest when it produces traceable outputs: what data sources were assessed, what rules or baselines were applied, how exceptions were handled, and what human oversight occurred.

If designed properly, this approach changes the cadence of compliance. Instead of audit preparation becoming an emergency mobilization, evidence becomes continuously organized and ready. Instead of compliance being a lagging indicator of risk, it becomes a leading indicator, surfacing control degradation early enough to correct it before it becomes a reportable failure.

Automating Narrative Generation for Compliance Reports

Even when evidence is strong, compliance can still fail if reporting is weak. Auditors and regulators do not only want telemetry. They want coherent explanations: how controls work, how they are monitored, and why the organization believes they are effective. Compliance reporting is therefore a translation function. It translates technical reality into structured assurance narratives that satisfy specific standards and regulatory expectations.

At scale, narrative generation becomes a major bottleneck because it requires assembling information from many sources and expressing it consistently across hundreds of controls and multiple frameworks. Human-authored narratives often vary by writer, business unit, and timing. They may be accurate but incomplete, or complete but inconsistent, or consistent but disconnected from evidence. AI can materially improve reporting if it is used as a disciplined drafting system anchored to validated evidence rather than as a free-form writing engine.

A strong AI narrative workflow begins with constraints. The system must know which framework is being addressed, what the control intent is, what evidence has been validated, what scope applies, and what language style is required. AI then generates a narrative that explains the control's design, its operational mechanism, how it is monitored, and what evidence supports that claim. The narrative is strongest when it includes the "how" and not only the "what," because auditors test operational reality. They want to know who reviews, how often, using what tools, what escalation occurs, and what corrective mechanisms exist when deviations are found.

This automation creates meaningful advantages, but only when paired with human oversight. Compliance reporting is not only a documentation exercise; it is a commitment. Therefore, AI-generated narratives must be reviewed by accountable owners who validate accuracy and context. The best model is not full autonomy but accelerated drafting. AI produces the first version that is evidence-aligned and framework-aware. Human reviewers finalize it, adjust for business nuance, and sign off. That sign-off becomes part of the audit trail, reinforcing accountability.

Narrative automation also improves the organization's internal operational clarity. When reports are generated consistently, gaps become more visible. If a narrative cannot be generated because evidence is missing, or if the evidence chain is weak, that deficiency becomes a remediation target rather than a surprise during an audit. Over time, reporting automation therefore becomes a feedback mechanism that forces improvement in evidence quality and control implementation.

Aligning AI Practices with Frameworks NIST SP and ISO

Framework alignment is where many AI compliance programs become fragile. Organizations often treat frameworks as checklists, mapping AI components to broad control families without verifying how AI changes the meaning of those controls. The more defensible approach is to treat frameworks as intent statements: they describe the outcomes compliance must ensure. AI must then be governed so those outcomes remain true even as systems evolve.

When mapping AI practices to standards such as the NIST Special Publications and ISO-based information security management systems, the organization must expand the control interpretation to include AI lifecycle realities. Access control is no longer only about who can log into systems; it includes those who can access training datasets, who can modify models, who can change inference endpoints, and who can influence feature pipelines. Audit and accountability are no longer only about system logs; it includes model versioning, training provenance, evaluation results, bias testing outputs, and decision trace evidence sufficient for audits. Risk assessment expands to include AI-specific threats such as model manipulation, data poisoning, adversarial inputs, and drift that changes compliance behavior. Configuration management expands to include

200

model configuration, inference settings, prompt governance for generative systems, and boundary conditions that limit inappropriate outputs.

ISO-aligned programs strengthen this effort because they emphasize management systems rather than single control checks. In a mature information security management system, AI systems are treated as assets with defined owners, classifications, risk treatment plans, operational monitoring, incident response integration, and continual improvement loops. Vendor and supplier governance becomes critical because many AI capabilities are sourced externally. Compliance must include contractual controls, transparency expectations, security requirements for data handling, and breach notification obligations. In other words, alignment is not achieved by stating that AI supports a framework. It is achieved by ensuring the framework's governance mechanisms fully include AI.

This is also the point where organizations must acknowledge that compliance is not only technical. It is socio-technical. If an AI system influences decisions affecting customers or employees, the compliance posture must include ethical and procedural safeguards. That includes oversight committees, escalation pathways, documentation practices, and fairness monitoring. These governance mechanisms are not "extra." They are now part of what it means to operate compliant systems in the AI era.

AI-Driven Continuous Compliance Monitoring

Continuous compliance is the natural endpoint of AI-assisted compliance at scale, but it must be designed as a governance capability rather than a surveillance capability. The purpose is not to generate constant alerts. The purpose is to sustain a reliable, current view of whether controls remain within acceptable boundaries, and to detect early indicators of drift before they become failures.

AI-driven monitoring becomes powerful when it is built on compliance signals. Compliance signals are measurable indicators that a control is operating: configuration posture against baseline, patch latency against thresholds, access events against policy, encryption coverage against standards, data retention against rules, and model behavior against approved performance and fairness profiles. AI systems can observe these signals continuously across complex environments, correlating deviations and prioritizing risk based on scope, impact, and historical patterns.

The compliance value is not simply speed. It is precision. Instead of discovering noncompliance during an audit window, the organization sees it as it emerges. Instead of guessing where risk is rising, the organization observes control degradation trends. Instead of relying on periodic attestations, the organization maintains operational proof.

This model also improves evidence quality. When monitoring is continuous and logs are immutable, the audit trail becomes richer and more credible. Evidence is not assembled after the fact; it is produced as a byproduct of governance. That reduces the temptation to rationalize gaps through documentation retrofits, which is one of the most common causes of audit breakdown under pressure.

However, continuous monitoring must preserve human accountability. AI can detect and surface deviations, but humans determine disposition. Some deviations reflect true failures; others reflect legitimate change. An effective model therefore includes human-in-the-loop review for ambiguous cases, defined thresholds for automated ticket creation, and escalation paths for repeat violations. Continuous compliance is not an algorithmic replacement for governance. It is governance with better instrumentation.

Board-Ready Brief

Compliance at Scale with AI Assistance: Continuous Assurance, Defensibility, and Governance Integrity

Executive Context

Regulatory compliance has shifted from a periodic validation exercise to a continuous operational obligation. As artificial intelligence becomes embedded across security operations, analytics, and decision workflows, the compliance surface expands rapidly and changes constantly. Traditional audit models built around point in time evidence collection and manual reporting cannot keep pace with AI driven environments where models evolve, data flows multiply, and automated decisions influence outcomes subject to regulatory and ethical scrutiny. The resulting risk is not simply noncompliance, but the inability to demonstrate compliance when it matters most.

This chapter frames AI assisted compliance as a governance capability rather than a reporting convenience. It explains how AI can be used to sustain defensible, auditable, and continuously validated compliance at enterprise scale while preserving human accountability and board level assurance.

Why Compliance at Scale Is Now a Board Level Issue

AI fundamentally alters the compliance equation by introducing systems that learn, adapt, and operate with partial autonomy. A control environment that appears compliant at audit time may drift silently as models retrain, configurations change, or data sources evolve. This creates a new category of exposure known as compliance drift. Without continuous validation, organizations risk unknowingly operating outside regulatory bounds even while maintaining formal policies and controls.

At the same time, regulators increasingly expect explainability, traceability, and accountability for automated decisions, particularly those affecting individuals, financial outcomes, or critical services. When AI systems cannot be clearly mapped to requirements, controls, evidence, and narrative justification, the organization's compliance posture becomes fragile. This elevates compliance from a technical or legal concern to a governance responsibility requiring sustained board oversight.

What the Board Should Expect from a Mature AI Assisted Compliance Model

Boards should expect compliance to function as a continuously monitored assurance system rather than a periodic audit artifact. A mature AI assisted model demonstrates how regulatory requirements are mapped to controls, how controls are continuously validated through telemetry and configuration data, and how deviations are detected and corrected before becoming findings or incidents. AI should reduce the time between control failure and remediation, not merely accelerate report production.

Leadership should also expect audit defensibility. Evidence should be traceable, time stamped, and reproducible. AI generated compliance narratives should be constrained by validated evidence and reviewed by accountable owners. The goal is consistency, completeness, and clarity so that auditors, regulators, and executives can understand not only that controls exist, but how they operate and why they remain effective over time.

Governance, Accountability, and Risk Containment

AI assisted compliance introduces new governance obligations. Responsibility for compliance outcomes cannot be diffused across models, vendors, or automation layers. Clear ownership must exist for AI systems, including model lifecycle management, data governance, control mapping, and exception handling. Decisions influenced by AI, particularly those affecting access, privacy, or regulatory exposure, must remain human owned and auditable.

Equally important is governing the AI itself as a compliance relevant system. Training data provenance, access controls, logging integrity, model drift monitoring, and third party AI risk must all be addressed explicitly. When AI becomes part of the compliance machinery, it must be governed with the same rigor as the controls it helps validate. This discipline prevents automation from becoming a hidden source of regulatory risk.

Trust, Transparency, and Continuous Assurance

Trust in AI assisted compliance is built through transparency and repeatability. Stakeholders must be able to see how evidence is collected, how controls are evaluated, and how narratives are generated. Continuous compliance monitoring shifts assurance from reactive explanation to proactive prevention, allowing leadership to address gaps before they escalate into findings, fines, or reputational damage.

When implemented with discipline, AI strengthens confidence rather than eroding it. Compliance teams move from manual data gathering to strategic risk oversight. Audit cycles shorten, findings decline, and the organization gains the ability to innovate without outpacing governance.

Board Oversight Questions

- How does the organization detect and correct compliance drift in AI driven systems
- Are AI generated compliance narratives constrained by validated evidence and human review
- Who owns accountability for AI influenced compliance outcomes

- How is explainability ensured for automated decisions subject to regulatory scrutiny
- Does continuous compliance monitoring reduce risk, or merely accelerate reporting

Compliance at scale is no longer achievable through manual processes or static audits. AI assisted compliance, when governed properly, becomes a strategic assurance capability that reduces regulatory exposure, supports responsible AI adoption, and provides the board with sustained confidence that innovation is not outpacing control.

Conclusion: Compliance at Scale

Compliance is changing because systems are changing. AI introduces opacity, evolution, and expanded data dependence, all of which strain compliance programs built for static environments. Organizations that attempt to manage AI-era compliance using the old model of periodic sampling and manual documentation will face growing fragility, rising costs, and increasing audit risk. The answer is not to abandon compliance rigor, but to strengthen it through better instrumentation and governance.

AI-assisted compliance at scale becomes viable when AI is used to map evidence continuously, validate controls through observable system behavior, generate narratives grounded in verified artifacts, and monitor compliance signals as a living discipline. Yet the chapter's central conclusion is that AI does not reduce the need for accountability. It increases it. Compliance remains a human obligation to prove that systems are lawful, secure, and ethically governed. AI can accelerate that proof, but only when implemented within a model that preserves explainability, traceability, and ownership. In the AI era, sustainable compliance was not achieved by producing more documentation. It is achieved by building an assurance system that stays true even as technology and risk continue to move.

Chapter 10: Workforce Enablement: AI Literacy and Skills

Artificial Intelligence is now embedded in day-to-day cybersecurity operations, shaping what teams see, how they prioritize risk, and how quickly they respond. This reality creates a simple constraint: AI tools only improve security outcomes when the workforce understands how to interpret and govern them. Chapter 10 establishes AI literacy as a baseline competency across the security organization and explains why role-specific training is required to translate literacy into operational performance. The chapter then addresses three practical enablement levers: prompt engineering as a security skill, quality control mechanisms that prevent unsupported AI outputs from making driving decisions, and continuous learning to keep pace with evolving models and adversary techniques. Workforce enablement is framed as a reliability and governance requirement rather than a training initiative. The objective is to build a security workforce that can collaborate with AI confidently, validate outputs responsibly, and apply AI-driven insights without surrendering accountability.

The Need for AI Literacy Across the Security Workforce

Artificial Intelligence is now embedded across security operations, driving detection, triage, prioritization, and recommendations that influence response actions. This reality means AI literacy must be distributed across the workforce, not concentrated only in data science or engineering teams. When practitioners treat AI as a black box, they either over-trust outputs and automate mistakes or distrust outputs and lose operational value. Baseline literacy equips teams to understand what AI can do, what it cannot do, and what evidence is required before acting. It also enables teams to interpret confidence signals and recognize when incomplete telemetry or biased patterns may be driving conclusions. AI literacy therefore functions as an operational control that improves reliability, speed, and defensibility.

In the Security Operations Center, AI literacy determines whether analysts become collaborators in intelligence-led defense or passive recipients of model outputs. AI-generated alerts and anomaly detections can be high value, but they can also produce false positives when baselines shift, or telemetry coverage is incomplete. Analysts who understand model reasoning and confidence cues can validate alerts efficiently and provide feedback that improves performance over time. Analysts without that literacy may waste effort chasing noise or may disregard important signals due to frustration or skepticism. AI literacy also supports consistent triage decisions because analysts learn how AI correlates signals and why prioritization changes. The result is a SOC that operates with higher confidence, lower fatigue, and stronger discipline around evidence.

AI literacy also matters for managers and leaders because AI outputs often describe risk probabilistically rather than deterministically. A forecasted increase in phishing likelihood or an elevated credential theft risk score requires interpretation and prioritization decisions grounded in business impact. Leaders need enough literacy to evaluate uncertainty, challenge assumptions, and translate predictions into pragmatic actions such as targeted controls, monitoring shifts, or training emphasis. Without this understanding, leaders can overcorrect toward low-likelihood threats or underinvest in high-impact risks that are not well captured by models. AI literacy bridges the gap between AI outputs and governance decisions that must be defensible. It ensures that strategic decisions are informed by AI without being dominated by it.

Ethical implications provide an additional reason AI literacy must be widespread, especially when models interact with identity, behavior, and insider risk signals. Bias can appear through training data, proxy variables, and uneven coverage across user groups or business units. Teams with literacy can test for uneven error rates, challenge questionable inferences, and require transparency for high-impact recommendations. This becomes critical when AI systems influence investigations, access restrictions, or disciplinary escalation pathways. A literate workforce protects the organization from operational harm and legitimacy erosion caused by unfair or opaque decision-making. Ethical literacy is therefore a trust requirement and a governance requirement in AI-enabled security operations.

Developing Role-Specific AI Training Programs

Role-specific training is the adoption engine because different cybersecurity roles interact with AI in different ways and carry different decision responsibilities. A generic AI course builds awareness, but it does not build competence in the workflows where AI changes outcomes. SOC analysts require hands-on training in interpreting AI triage outputs, understanding confidence scores, and recognizing the difference between anomaly signals and confirmed threat behaviors. Incident responders require training that focuses on AI-assisted evidence gathering, timeline reconstruction, and containment recommendations under time pressure. Security architects require training in AI integration patterns, data requirements, model risk, and resilience against manipulation and drift. Leaders require training that emphasizes governance, value measurement, and accountability boundaries between AI recommendation and execution.

For SOC analysts, the training goal is operational fluency rather than model development expertise. Analysts should learn how AI-driven correlation engines assemble context and why a low-severity event may be elevated into a critical incident when linked to attack chains. Analysts should practice validating AI outputs by checking telemetry coverage, reviewing underlying signals, and confirming that enrichment data is relevant and current. Training should include exercises that require analysts to label outcomes, document rationale, and feed structured feedback back into model tuning processes. This builds a human-in-the-loop learning posture and increases analyst trust because the system improves visibly. The objective is to reduce noise, speed triage, and improve decision consistency without lowering investigative rigor.

For incident responders, training should focus on AI as an accelerator rather than an authority. Responders should learn where AI can automate repetitive tasks such as artifact collection, preliminary scoping, and indicator expansion, while preserving human judgment for uncertain scenarios. Training should include simulations where AI produces incomplete or partially incorrect hypotheses, requiring responders to validate and correct the narrative using evidence. Responders should also practice using AI to generate structured communications such as executive summaries, containment status updates, and post-incident action lists, while applying strict validation standards. This approach builds speed without sacrificing accuracy, especially when time pressure creates vulnerability to overreliance. The outcome is faster containment and clearer incident documentation.

For architects and engineering leaders, training must address the operational dependencies AI introduces, including data pipelines, access control, and model lifecycle management. Architects should learn how to evaluate vendors and tools based on telemetry coverage, explainability, audit support, and integration with existing platforms. Training should include model risk topics such as drift, poisoning, and prompt injection risks in generative tooling, along with compensating controls that preserve reliability. Architects should also understand privacy and retention design choices because SOC AI systems often touch sensitive identity and behavioral signals. This enables architectures that are secure, governed, and resilient under real-world conditions. The outcome is AI adoption that scales without creating hidden fragility.

Mastering Prompt Engineering for Security Applications

Prompt engineering is a practical security skill because generative AI systems produce reliable outputs only when the user provides clear context, constraints, and validation expectations. Poor prompts produce generic responses, invented details, and misleading certainty, which can pollute incident records and mislead decision-making. Strong prompts specify the task, scope, data inputs, required evidence, and output format, and they require the model to disclose uncertainty when information is missing. In security operations, prompt quality affects investigative summaries, playbook steps, detection hypothesis generation, executive reporting, and policy drafting. Prompt engineering also reduces workflow friction because it improves clarity and repeatability of AI-assisted tasks. Organizations should treat prompt patterns like playbooks: reusable, reviewed, and governed.

In SOC workflows, prompts should be designed to produce evidence-driven summaries rather than narrative speculation. A strong investigation prompt asks for correlated events, timelines, affected assets, and confidence rationale tied to observable signals. It also demands citations to specific telemetry sources such as endpoint events, identity logs, and network connections, even if the system only provides reference IDs internally. Analysts should be trained to request alternative hypotheses and to ask the model to list what it cannot know from the available data. This reduces

211

overconfidence and encourages verification. A disciplined prompt practice turns generative AI into an analyst amplifier rather than a source of noise.

Prompt engineering also supports leadership communications by structuring AI outputs into board-ready language without sacrificing accuracy. Executives need concise summaries that describe impact, actions taken, residual risk, and next steps, not model internals or raw technical artifacts. Prompts should require clear separation between confirmed facts, inferred judgments, and recommended actions, with uncertainty statements when evidence is partial. This distinction prevents AI-generated reports from overstating certainty and protects the organization during audits or legal reviews. Over time, organizations should develop a prompt library aligned to incident classes, operating rhythms, and reporting requirements. This makes AI usage consistent, trainable, and measurable.

Implementing Quality Control Mechanisms for AI-Generated Outputs

AI-generated security outputs must be treated as untrusted until validated because model errors can create operational harm at machine speed. Quality control begins with classification of AI outputs by impact, where high-impact decisions require stronger evidence and mandatory human approval. Automated enrichment and summarization can be largely automated if output remains observable and reversible, but actions such as isolation, lockout, or business disruption require explicit human gating. Quality control also requires cross-validation against authoritative sources such as telemetry systems, ticketing records, and verified threat intelligence. The organization should define acceptance criteria including minimum evidence thresholds, confidence requirements, and escalation triggers. This ensures that AI supports decisions rather than silently making them.

Human review checkpoints should be embedded in workflows where AI outputs affect containment, external reporting, compliance claims, or employee-facing outcomes. Review should emphasize evidence quality, completeness of data coverage, and whether alternative explanations were considered. Reviewers should be trained to detect hallucinated specificity, where AI provides detailed but unsupported claims, especially in timelines and attribution. Quality control should also include structured reviewer feedback captured for continuous improvement of prompts and model configuration. When feedback is not captured, the organization repeats errors and loses the learning benefit. A disciplined review loop converts AI from a brittle tool into a continuously improving capability.

Quality control must also account for drift and changing baselines that can degrade model performance over time. The SOC should monitor false positive rates, false negative indicators, and shifts in alert distribution to identify when models are losing relevance. Periodic red-team style testing and scenario evaluation should be used to validate that AI-assisted detection remains effective against evolving tactics. Governance should require documented fallbacks when models degrade, including reverting to known reliable detection rules and adjusting automation thresholds. These controls preserve resilience and prevent AI dependency from becoming an operational single point of failure. The objective is to sustain reliability under changing conditions.

Fostering Continuous Learning and Adaptation

Continuous learning is required because AI tools, adversary tactics, and organizational environments change faster than traditional training cycles. Enablement must be structured as a recurring program with refreshers, labs, simulations, and role-based capability progression. Knowledge sharing should be operationalized through internal briefings, prompt libraries, case retrospectives, and controlled experimentation in sandbox environments. Teams should be encouraged to treat AI failures as learning events, capturing what went wrong and updating processes rather than assigning blame. This creates a culture where AI is used actively and responsibly rather than passively or fearfully. Continuous learning keeps humans aligned with AI the same way monitoring keeps models aligned with reality.

Adaptation also requires investment in hybrid skills that bridge security practice and AI-enabled workflows. Analysts need data literacy to interpret model signals, responders need disciplined verification habits under pressure, and leaders need governance literacy to manage risk and accountability. Organizations should establish "AI champions" in each security function who help translate new capabilities into practical workflows and train peers. Learning should be measured using operational outcomes such as reduced triage time, fewer noisy escalations, improved investigation quality, and clearer executive reporting. This ties enablement to performance rather than to course completion. The goal is an adaptable workforce that improves as fast as the tools evolve.

Board-Ready Brief

Executive Overview

Artificial Intelligence is now embedded across cybersecurity operations, from detection and response workflows to vulnerability prioritization and policy validation. This change shifts cybersecurity performance from being primarily tool-driven to being workforce-enabled, because the quality of outcomes depends on how well practitioners interpret, govern, and operationalize AI outputs. AI literacy is therefore not an optional training initiative but a prerequisite for safe adoption, reliable decisions, and defensible governance. A workforce that treats AI as a black box will either overtrust it and automate harm, or distrust it and fail to capture value.

Chapter 10 frames AI enablement as a strategic control: building shared literacy, role-specific competencies, prompt discipline, and quality assurance mechanisms. It also emphasizes continuous learning as the only sustainable posture as AI systems and adversary techniques evolve.

Why AI Literacy Must Be Enterprise-Wide in Security

AI-enabled security tools influence prioritization, triage, and containment decisions, meaning they can change operational outcomes even when humans remain in the loop. Without baseline literacy, analysts may accept AI outputs as authoritative, or dismiss them without understanding model confidence, data coverage, or limitations. Security leaders also require literacy to translate probabilistic AI insights into risk decisions that align with business impact and regulatory obligations. AI literacy reduces error and improves speed because practitioners can ask better questions, validate outputs efficiently, and provide corrective feedback that improves models over time. This literacy also supports trust because teams can explain how AI contributed to decisions and can defend those decisions in audits or post-incident reviews. In practical terms, AI literacy is a workforce control that directly improves mean time to triage, reduces false positives, and strengthens decision consistency.

Role-Specific Training as the Adoption Engine

A single training track cannot serve a security organization because the SOC analyst, incident responder, security architect, and security leader interact with AI differently. Analysts need practical competence in interpreting AI alerts, confidence signals, and enrichment context, while learning how to label outcomes and improve model feedback loops. Incident responders must understand where AI accelerates evidence gathering, timeline reconstruction, and containment recommendations, and where human judgment must override automation. Architects must evaluate AI tooling, data pipelines, integration patterns, and model risks such as drift, poisoning, and dependency on incomplete telemetry. Leaders must focus on governance, value realization, and the policy boundary between AI-recommended and AI-executed actions. Role-specific training translates AI literacy into operational competence, reducing adoption friction and preventing unsafe reliance on AI outputs.

Prompt Engineering as a Security Skill

Generative AI systems behave like force multipliers only when users provide structured instructions, constraints, and validation expectations. Prompt discipline becomes especially important when generating investigative summaries, incident reports, detection logic drafts, and executive communications. Poor prompts produce vague responses, hallucinated details, and misleading certainty, which can contaminate incident response timelines and decision-making. Strong prompts define context, scope, constraints, required evidence, and output format, while also requiring uncertainty disclosure when data is missing. Prompt engineering is not a novelty; it is a practical competency that improves analyst speed and report quality, and reduces risk created by AI overconfidence. Organizations should treat prompt patterns as reusable operational assets, governed like playbooks and detection rules.

Quality Control for AI-Generated Outputs

AI-generated outputs must be treated as untrusted until validated, especially when they influence containment, communications, compliance, or disciplinary decisions. Quality control requires a layered approach: automated checks where feasible, structured human review at decision points, and cross-validation against authoritative data sources. The organization should define acceptance criteria for AI-assisted conclusions, including minimum evidence requirements, confidence thresholds, and escalation triggers.

High-impact actions must require human approval and must preserve audit trails showing what the AI recommended and why. These mechanisms prevent error propagation and reduce the chance that AI outputs become "institutional truth" without proof. Quality control is also a learning engine because reviewer feedback should be captured and used to improve prompts, workflows, and model configurations.

Continuous Learning and Workforce Adaptation

AI tools, models, and threats evolve continuously, meaning enablement cannot be a one-time training event. Security teams need a continuous learning loop: updated training, internal knowledge sharing, controlled experimentation, and measurement tied to operational outcomes. A structured enablement program should include labs, simulation-based training, periodic refreshers, and role-based capability progression with clear proficiency expectations. Organizations should also prepare for workforce shifts, including the rise of hybrid roles such as detection engineers who collaborate with AI, analysts who interpret model outputs, and security leaders who govern AI risk. Continuous learning reduces drift in human competence the same way model monitoring reduces drift in AI performance. This is the sustainability layer that makes AI adoption durable rather than fragile.

Board-Level Oversight Questions

- What decisions are AI-influenced versus AI-executed, and where is human approval mandatory?
- How are security teams trained to interpret AI confidence, data limitations, and failure modes?
- What quality control gates prevent AI hallucinations or unsupported claims from entering incident records or executive reporting?

- How do we measure workforce enablement outcomes: reduced false positives, faster triage, improved consistency, fewer escalations driven by noise?
- What is the plan for continuous training updates as models and adversary techniques evolve?
- Who owns AI risk in the security workforce, and how is accountability documented?

Conclusion for the Board

AI capability in cybersecurity is constrained by workforce readiness, not by tool availability. Chapter 10 positions enablement as a security control that supports reliability, defensibility, and value realization. Organizations that invest in role-specific training, prompt discipline, and quality assurance build teams that can safely operationalize AI and improve outcomes without introducing new risk. Continuous learning is the sustaining mechanism that keeps both tools and humans aligned with a rapidly changing threat environment. The board should treat AI enablement as a strategic resilience investment with measurable operational returns and explicit governance requirements.

Board Oversight Appendix: AI Workforce Enablement, Metrics, and Roadmap

Appendix A: Executive Oversight Metrics (KPIs)

Effective oversight of AI workforce enablement requires metrics that reflect operational outcomes rather than training activity alone. Boards should expect reporting that links AI literacy and skills development to measurable improvements in security performance and decision reliability. Key indicators include reductions in false positives, improvements in mean time to triage and containment, and consistency of response actions across teams and shifts. Additional metrics should track how often AI outputs require correction, escalation, or override, as this reveals both model maturity and workforce judgment quality. Training completion rates are necessary but insufficient and should be treated as input metrics rather than outcomes. The board should prioritize metrics that demonstrate safer, faster, and more defensible decisions driven by AI-enabled workflows.

Recommended workforce-aligned KPIs include analyst alert validation accuracy, average investigation time per AI-prioritized incident, and the percentage of AI-assisted reports requiring substantive revision before executive distribution. Leadership should also monitor the frequency of AI-related operational incidents, such as incorrect automated actions or misleading summaries, and the time required to detect and correct them. Feedback loop effectiveness can be measured by the percentage of AI outputs that receive structured human feedback and result in model or workflow updates. Ethical assurance metrics should include bias testing results, audit findings, and documented cases where AI recommendations were modified due to fairness or governance concerns. Together, these indicators provide visibility into whether AI enablement is improving outcomes without introducing hidden risk.

Appendix B: AI Workforce Enablement Maturity Model

AI workforce enablement progresses through identifiable maturity stages that reflect how deeply literacy, skills, and governance are embedded. At the Foundational stage, AI tools are in use but understanding is uneven, prompts are ad hoc, and outputs are trusted or dismissed without consistent validation. At the Operational stage, role-specific training exists, prompt standards are emerging, and human review checkpoints are defined for high-impact actions. The Integrated stage is marked by consistent AI literacy across roles, documented prompt libraries, formal quality control criteria, and measurable performance improvements tied to

AI usage. At the Adaptive stage, continuous learning loops, model feedback mechanisms, and governance reviews are embedded into daily operations. Boards should expect organizations deploying AI at scale to operate at least at the Integrated stage to ensure reliability and defensibility.

Maturity assessment should include both workforce capability and governance alignment. Indicators of higher maturity include consistent interpretation of AI confidence signals, disciplined escalation practices, and routine validation of AI-generated outputs before action. Lower maturity environments often show overreliance on automation, inconsistent triage decisions, and unclear accountability when AI-driven errors occur. The maturity model provides a structured lens for evaluating progress without focusing solely on tool deployment. It also allows leadership to identify where investment should be directed, whether in training, governance, or workflow redesign. Boards should request periodic maturity assessments as part of AI risk oversight.

Appendix C: 12-Month AI Workforce Enablement Roadmap

A structured roadmap enables disciplined progress while reducing disruption to ongoing security operations. In the first quarter, organizations should establish baseline AI literacy expectations, identify role-specific skill gaps, and define governance boundaries for AI-assisted versus AI-executed actions. This phase should include foundational training, initial prompt standards, and clear quality control checkpoints for high-impact decisions. The second quarter should focus on role-specific training tracks, hands-on labs, and the development of approved prompt libraries aligned to SOC, incident response, and leadership workflows. Measurement should begin in this phase, with baseline KPIs captured for comparison.

In the third quarter, organizations should operationalize feedback loops, integrating analyst and reviewer input into prompt refinement and model configuration. Governance reviews should validate that AI usage aligns with policy, ethical expectations, and regulatory obligations. The fourth quarter should emphasize optimization and sustainability, including refresher training, drift monitoring, and scenario-based exercises that test human-AI collaboration under stress. By the end of twelve months, AI literacy should be normalized across the workforce, quality control mechanisms should be routine, and leadership should have clear evidence of improved outcomes. This roadmap supports steady progress without forcing premature automation or unsafe acceleration.

Appendix D: Board-Level Accountability and Assurance

Boards play a critical role in ensuring that AI workforce enablement remains aligned with organizational risk appetite and values. Oversight should confirm that accountability for AI-related decisions is clearly assigned and documented, particularly when AI influences containment, access, or employee-related actions. Boards should require assurance that AI-generated outputs are auditable and that decision rationales can be reconstructed after incidents. Regular briefings should include summaries of AI-related errors, corrective actions taken, and lessons learned, reinforcing a culture of transparency. The board should also confirm that workforce enablement keeps pace with AI capability expansion, preventing gaps between what tools can do and what people understand. Effective oversight ensures AI strengthens security operations without eroding trust, control, or responsibility.

Conclusion: AI Literacy and Skills

AI adoption in cybersecurity succeeds when enablement is treated as a workforce and governance transformation rather than a tooling upgrade. AI literacy reduces risk by preventing blind trust and uninformed skepticism, enabling practitioners to interpret model confidence, validate conclusions, and provide feedback that improves performance over time. Role-specific training ensures that analysts, responders, architects, and leaders develop the competencies required for their decisions and workflows, while prompt engineering improves precision and reduces ambiguity in AI-assisted tasks. Quality control mechanisms protect the organization by ensuring AI outputs are evidence-based, auditable, and gated appropriately for high-impact actions. Continuous learning keeps the workforce aligned with evolving threats and rapidly changing AI capabilities, preventing capability drift and unsafe reliance. When these elements mature together, the organization gains faster response, better prioritization, and stronger defensibility without compromising human accountability

Chapter 11: Architecting for Resilience: The Human-AI Ecosystem

Resilience in cybersecurity has traditionally been defined by system recovery and operational continuity, but artificial intelligence fundamentally changes what resilience requires and how it must be designed. AI-enabled security systems introduce adaptive capabilities that can dramatically improve detection and response, while simultaneously creating new failure modes, including silent degradation, adversarial manipulation, and data-driven distortion. Chapter 11 reframes resilience as an ecosystem property that emerges from the interaction of humans, AI systems, data pipelines, and governance structures rather than from technology alone. It examines how organizations can architect security environments that remain trustworthy and effective even when AI systems fail, degrade, or behave unpredictably. The chapter emphasizes that resilience is not the absence of disruption, but the ability to sustain sound decision-making under stress. The goal is to equip leaders with a framework for designing AI-enabled security architectures that adapt, recover, and strengthen over time.

Defining Organizational Resilience in the AI Context

Organizational resilience in the age of artificial intelligence extends beyond traditional business continuity and disaster recovery planning. While those foundations remain necessary, resilience in an AI-enabled security environment requires the capacity to adapt as threats, systems, and decision logic continuously evolve. Artificial intelligence introduces dynamic risk because both adversaries and defenders can leverage learning systems that change behavior over time. Defensive AI systems themselves may degrade, drift, or be targeted for manipulation, creating new categories of operational failure. Resilience therefore must be understood as the organization's ability to anticipate AI-amplified disruption, absorb operational shock, respond with coordinated human-AI decision-making, and recover in ways that improve future performance. This framing positions resilience as an ongoing capability rather than a static end state.

The AI context fundamentally reshapes the nature of disruption by accelerating scale, speed, and complexity. AI-enabled attackers can automate reconnaissance, generate polymorphic malware, and adapt tactics faster than traditional defensive cycles. At the same time, AI-driven defensive systems can fail silently, misclassify threats, or produce misleading confidence signals. A compromised or degraded AI security control may not visibly fail, yet still undermine situational awareness. Resilience therefore requires explicit consideration of AI-specific failure modes, including adversarial manipulation, model drift, and dependency on corrupted data. Organizations that do not account for these risks may experience systemic blind spots rather than isolated outages.

Anticipation becomes a core pillar of resilience in this environment. Organizations must leverage threat intelligence, predictive analytics, and adversary modeling to identify emerging AI-driven attack patterns before they materialize operationally. This includes monitoring for techniques that exploit AI systems directly, such as data poisoning or evasion attacks. Anticipation also requires understanding how geopolitical events, technology adoption trends, and regulatory shifts influence adversary behavior. By integrating AI-driven forecasting with human analysis, organizations can prioritize defensive investments proactively. This anticipatory posture reduces the likelihood that AI-enabled threats will overwhelm defensive capacity.

Withstanding disruption requires designing AI-enabled security architectures for robustness rather than perfection.

AI systems must be treated as critical infrastructure components that require redundancy, segmentation, and controlled failure modes. Defensive capabilities should not collapse when a single model, data pipeline, or automation layer fails. Instead, organizations must ensure that core security functions continue operating at a reduced but reliable level. This includes maintaining fallback detection mechanisms and preserving human visibility into security conditions. Robustness ensures that AI enhances resilience rather than becoming a single point of failure.

Response in an AI-driven incident must balance speed with judgment. AI systems can rapidly surface signals and automate initial actions, but human oversight remains essential when consequences affect business operations, legal exposure, or safety. Security teams must be trained to recognize when AI outputs are reliable and when they require skepticism. Human-AI teaming enables rapid containment while preserving contextual decision-making. Effective response depends on clear escalation paths and the ability to override automation without friction. This coordinated approach ensures that AI accelerates response without undermining accountability.

Recovery in the AI context is inseparable from learning. Post-incident analysis must evaluate not only what the adversary did, but how AI systems performed under stress. Organizations should assess where AI succeeded, where it failed, and how human intervention influenced outcomes. These insights must feed directly into model retraining, process refinement, and governance adjustments. Recovery therefore becomes a mechanism for strengthening future resilience rather than merely restoring prior conditions. Over time, this learning cycle enables the organization to become more adaptive and resistant to AI-amplified threats.

Designing AI Systems for Robustness and Fault Tolerance

Robust AI system design begins with the assumption that failure is inevitable rather than exceptional. AI models operate within complex data environments and are subject to both technical and adversarial disruption. Resilient architectures therefore require redundancy at multiple layers, including models, data pipelines, and execution environments. Multiple detection approaches reduce dependence on any single analytic technique. This diversity limits the impact of bias, drift, or targeted manipulation. Robust design ensures continuity of security insight even when individual components degrade.

Model-level redundancy is particularly important in security applications. Deploying diverse models trained on different datasets and architectures reduces systemic risk. If one model becomes unreliable, others can provide compensating signals. Cross-model validation also enables detection of anomalous AI behavior. Discrepancies between model outputs can trigger investigation rather than silent failure. This approach treats AI outputs as probabilistic inputs rather than absolute truth.

Fault tolerance must extend beyond models to data ingestion and processing pipelines. AI systems are only as reliable as the data they consume. Corrupted, delayed, or manipulated data streams can undermine detection accuracy. Resilient architectures therefore include integrity checks, validation controls, and alternate data sources. Monitoring data behavior itself becomes a defensive capability. Early detection of anomalous data patterns protects downstream AI performance.

Failover mechanisms are essential for maintaining operational continuity. AI-driven security functions should automatically transition to backup systems when failures occur. These transitions must be rapid and visible to operators. Standby systems must remain synchronized to avoid operational gaps. The choice between hot and cold failover depends on mission criticality. Effective failover ensures that security posture degrades gracefully rather than collapsing.

Graceful degradation is a defining feature of resilient AI systems. When advanced analytics fail, systems should revert to simpler detection or human-driven workflows. Reduced capability is preferable to false confidence. This requires explicit prioritization of security functions and pre-defined fallback behaviors. Graceful degradation preserves trust and situational awareness under stress. It ensures that AI failure does not equate to security blindness.

Human oversight remains a structural requirement for fault tolerance. Analysts must understand AI limitations and failure indicators. Interfaces should support rapid interpretation and intervention. Training must emphasize recognizing degraded AI behavior as well as adversarial exploitation. Human judgment provides the final safeguard against cascading failure. Resilient design integrates people as active components, not passive users.

The Role of Data Provenance and Integrity in Resilience

Data provenance and integrity form the foundation of trustworthy AI-driven security operations. Without confidence in data origin and handling, AI outputs cannot be reliably interpreted or defended. Provenance enables traceability across the AI lifecycle, from ingestion to inference. This visibility is essential when diagnosing unexpected model behavior. It allows teams to identify whether errors originate from data sources, processing logic, or adversarial manipulation. Provenance transforms AI from a black box into an auditable system.

Adversarial data poisoning represents a significant resilience risk. Attackers can manipulate training or inference data to bias AI outcomes. These manipulations may be subtle and persist over time. Without provenance, such attacks can remain undetected until damage is severe. Provenance controls allow organizations to audit data flows and isolate compromised datasets. This capability is essential for recovery and retraining.

Data integrity ensures that information remains accurate and unaltered throughout its lifecycle. Cryptographic validation, access control, and versioning protect against accidental corruption and malicious tampering. Integrity failures can propagate error across AI systems rapidly. Defensive integrity controls therefore function as resilience safeguards. They prevent AI systems from amplifying corrupted inputs into flawed decisions.

Human oversight complements technical data controls. Analysts with domain expertise often detect inconsistencies that automated systems miss. Providing visibility into data lineage empowers humans to challenge AI conclusions when warranted. This collaboration strengthens trust and accountability. Provenance and integrity thus support both technical resilience and human confidence. Together, they enable defensible AI-driven decision-making.

Integrating Human Judgement with AI for Adaptive Responses

Human judgment is the adaptive control layer within the AI-enabled security ecosystem. AI excels at pattern recognition and scale, but humans interpret intent, context, and consequence. This distinction is critical when responding to novel or ambiguous threats. AI may flag anomalies without understanding business significance. Humans bridge that gap by applying organizational knowledge and strategic reasoning. Adaptive response emerges from this partnership.

Human oversight is essential when automated actions carry operational risk. Decisions such as isolating systems or revoking access may have significant business impact. AI can recommend actions, but humans must authorize execution in high-impact scenarios. This preserves accountability and aligns response with risk tolerance. Clear decision boundaries prevent over-automation. Human judgment ensures proportional response.

Explainability strengthens human-AI collaboration. Analysts must understand why AI produced a recommendation. Transparent reasoning supports trust and validation. When explanations are unclear, analysts can investigate further rather than acting blindly. Explainability also supports learning by highlighting where AI logic diverges from human expectation. This feedback improves future performance.

Adaptive learning depends on structured human feedback. Analyst decisions should inform model refinement and governance adjustments. Overrides and corrections become training signals rather than failures. This continuous loop allows AI systems to evolve alongside adversary tactics. Human judgment thus shapes AI behavior over time. Resilience emerges through shared learning.

Testing and Validating the Resilient Human-AI Ecosystem

Resilience claims must be validated under realistic stress conditions. Testing must simulate adversarial behavior, AI failure, and operational disruption. This includes data corruption, model drift, and automation errors. Validation should assess both technical performance and human-AI coordination. Effective testing reveals weaknesses before adversaries exploit them. It converts assumptions into evidence.

Adversarial simulations are essential for testing detection and response capability. Red team exercises should challenge AI systems directly and indirectly. These scenarios reveal how well humans interpret AI outputs under pressure. Success depends on collaboration rather than automation alone. Testing must evaluate decision quality, not just detection speed. This ensures resilience under real-world conditions.

Component failure testing assesses fault tolerance and graceful degradation. Disabling AI modules or data feeds tests fallback mechanisms. Human analysts must compensate without losing situational awareness. Metrics such as detection continuity and response delay provide objective insight. These exercises expose hidden dependencies. They strengthen system design.

Trust calibration is a critical validation objective. Analysts must neither over-trust nor disregard AI outputs. Testing should include misleading AI signals to assess human skepticism. Successful teams question AI appropriately. This balance is essential for resilience. Validation ensures healthy human-AI dynamics.

Testing is not a one-time effort but a continuous discipline. Threats evolve, AI systems change, and organizational contexts shift. Regular validation ensures resilience keeps pace. Dedicated testing environments support experimentation and improvement. Continuous testing transforms resilience from theory into practice.

Board-Ready Brief

Architecting for Resilience: The Human-AI Ecosystem

Executive Overview

As artificial intelligence becomes embedded in cybersecurity operations, organizational resilience can no longer be measured solely by uptime, recovery time, or infrastructure

230

redundancy. AI introduces adaptive systems that learn, evolve, and sometimes fail in ways that are opaque and difficult to detect. Defensive AI systems can drift, misclassify threats, or become targets themselves, while AI-enabled adversaries can operate at unprecedented speed and scale. In this environment, resilience must be intentionally architected across people, processes, data, and AI systems as a unified ecosystem. Chapter 11 positions resilience as a continuous, adaptive capability that allows organizations to anticipate disruption, withstand AI-specific failure modes, respond with coordinated human-AI decision-making, and recover in ways that improve future performance.

Resilience in an AI-Enabled Security Environment

AI fundamentally alters the nature of cybersecurity disruption by accelerating change and increasing system complexity. Traditional resilience models assume relatively static systems and predictable failure modes, but AI-driven systems evolve continuously based on data and feedback. This creates new risks, including silent model degradation, biased decision-making, and adversarial manipulation of learning processes. Resilience in this context requires the ability to maintain decision integrity even when AI components perform imperfectly. Organizations must therefore design security architectures that preserve visibility, control, and accountability under AI stress conditions. Resilience becomes a property of the ecosystem rather than a function of individual tools.

Designing AI Systems for Robustness and Fault Tolerance

AI systems supporting security operations must be architected with failure as an expected condition rather than an anomaly. Robust design includes redundancy across models, architectures, and data sources to reduce dependence on any single analytic capability. Fault tolerance ensures that when AI components degrade or fail, core security functions continue operating through fallback mechanisms and human intervention. Graceful degradation is critical, allowing systems to revert to simpler detection or manual workflows instead of producing false confidence. Automated failover and continuous health monitoring further protect continuity. Together, these design principles prevent AI from becoming a single point of failure in the security architecture.

Data Provenance and Integrity as Resilience Controls

The effectiveness and trustworthiness of AI-driven security operations depend on the integrity and traceability of data. Data provenance provides visibility into where data originates, how it is transformed, and how it influences AI decisions. Without this visibility, diagnosing AI failures or adversarial manipulation becomes nearly impossible. Integrity controls such as cryptographic validation, access restriction, and dataset versioning protect against corruption and poisoning. These controls are not merely technical safeguards but resilience mechanisms that enable recovery and defensibility. Strong data governance ensures that AI amplifies insight rather than propagating error at scale.

Human Judgment as the Adaptive Control Layer

While AI excels at processing scale and detecting patterns, human judgment remains essential for interpreting intent, context, and consequence. High-impact security decisions often involve business tradeoffs, legal considerations, and ethical implications that AI cannot independently evaluate. Human oversight ensures that automated recommendations are validated and appropriately constrained. Explainable AI supports this partnership by enabling analysts to understand and challenge AI conclusions. Structured feedback loops allow human decisions to continuously refine AI behavior. This integration ensures adaptive response rather than rigid automation.

Testing and Validating the Human-AI Ecosystem

Resilience must be demonstrated through testing rather than assumed through design. Organizations should conduct adversarial simulations, AI failure scenarios, data corruption exercises, and human-AI coordination drills. These tests evaluate how well the ecosystem performs under stress, not just during normal operations. Metrics such as detection continuity, decision quality, and recovery effectiveness provide objective assurance. Validation should also assess trust calibration, ensuring analysts neither over-rely on nor disregard AI outputs. Continuous testing converts resilience from a theoretical goal into an operational reality.

Board Oversight Considerations

- How do we detect and respond when AI systems degrade or behave incorrectly?
- Which security decisions remain safe when AI is unavailable or unreliable?
- What data integrity controls protect AI training and inference pipelines?
- How is human oversight enforced for high-impact automated actions?
- How often is AI resilience tested under adversarial and failure conditions?
- Who is accountable for AI-related operational risk?

Executive Takeaway

AI increases defensive capability while simultaneously introducing new systemic risks. Organizations that architect resilience across the human-AI ecosystem gain adaptability, transparency, and control under stress. Those that treat AI as a black box risk silent failure and cascading disruption. Chapter 11 demonstrates that resilience is not achieved by eliminating failure, but by designing systems, workflows, and governance structures that absorb disruption and learn from

233

it. When resilience is engineered intentionally, AI becomes a durable force multiplier rather than a fragile dependency.

Conclusion: Human-AI Ecosystem

Resilience in an AI-enabled security environment cannot be achieved through technology alone. Artificial intelligence introduces powerful defensive capabilities, but it also reshapes the nature of risk by embedding learning systems, probabilistic decision-making, and automation into core security functions. These characteristics require organizations to rethink resilience as an ecosystem property that emerges from the interaction of AI systems, data integrity, human judgment, and governance structures. When any one of these elements is neglected, AI becomes a source of fragility rather than strength.

Architecting for resilience begins with acknowledging that AI systems will fail, drift, and be targeted by adversaries. Robust and fault-tolerant design ensures that such failures are visible, contained, and recoverable rather than silent and cascading. Data provenance and integrity provide the trust foundation necessary for diagnosing failures and restoring confidence in AI-driven decisions. Without these controls, recovery becomes guesswork, and learning is undermined. Resilience therefore depends as much on disciplined data management as on advanced analytics.

Human judgment remains the adaptive control layer within this ecosystem. While AI accelerates detection and response, humans provide context, ethical reasoning, and strategic prioritization under uncertainty. Effective resilience is demonstrated when human operators can confidently validate, challenge, and override AI outputs, when necessary, while also feeding insights back into continuous improvement cycles. This human–AI partnership ensures that speed does not replace accountability and that automation remains aligned with organizational intent.

Finally, resilience must be validated through continuous testing rather than assumed through design. Adversarial simulations, AI failure scenarios, and human–AI coordination exercises transform resilience from an abstract objective into an operational capability. These practices reveal weaknesses, recalibrate trust, and reinforce adaptive behaviors before real-world crises occur. When organizations intentionally design, govern, and test the human–AI ecosystem, resilience becomes a sustained advantage rather than a reactive aspiration. In an era of AI-amplified threats, this architectural approach is what allows security programs not only to endure disruption, but to emerge stronger from it.

Chapter 12: Communicating AI Strategy to Executive Leadership

Artificial intelligence in cybersecurity succeeds or fails as a leadership story before it succeeds or fails as a technical system. Executive leadership and boards do not fund models, pipelines, or platforms. Executive leadership funds outcomes: reduced enterprise risk, protected revenue, operational resilience, compliance confidence, and sustained customer trust. This chapter provides a structured approach for security leaders to translate AI capability into business value, present risk with credibility, measure success with executive-level metrics, and build the trust required to sustain multi-quarter AI programs. The objective is to create communication that is not persuasive in tone alone, but defensible in substance, grounded in measurable results, and aligned to the organization's strategic priorities.

Bridging the Gap - Technical AI Concepts to Business Value

The core challenge in any AI security initiative is not the complexity of the technology. The core challenge is relevance. Executive leadership operates in a governance environment shaped by earnings pressure, fiduciary responsibilities, brand expectations, regulatory scrutiny, and strategic growth targets. Security leaders must therefore translate AI from "how it works" into "why it matters," and from "what it can do" into "what it will change." This translation is not a simplification. It is a reframing that makes the initiative legible to decision-makers who fund enterprise outcomes, not technical novelty.

This translation begins by anchoring AI initiatives to specific business exposures and strategic objectives. An AI detection model should not be introduced as a machine learning advancement. It should be introduced as a lever that reduces business email compromise risk, limits fraud loss exposure,

236

and increases continuity of operations. An AI-enabled threat intelligence capability should not be described through ingestion rates, entity extraction methods, or model architecture. It should be framed as earlier identification of material threats, faster executive decision cycles during incidents, and reduced probability of disruptive outages that degrade market confidence.

Risk reduction is one of the clearest executive value pathways, and AI must be framed as a measurable risk control rather than a tool upgrade. AI reduces risk when it lowers the likelihood of material incidents, reduces the severity of impact when incidents occur, and improves recovery time when disruption happens. AI also reduces risk by reducing human error exposure in repetitive security processes. When automation improves triage consistency, validates asset context faster, and accelerates containment, the enterprise reduces the probability of fatigue-based failures, delayed response, and uncontrolled incident expansion.

Cost savings and efficiency gains must be treated carefully to remain credible. Executives will correctly challenge inflated productivity claims if the organization lacks the maturity process to operationalize AI. The strongest approach is to frame cost value as a combination of operational efficiency and cost avoidance. Operational efficiency is measured through reduced time to detect and respond, reduced false positives, and improved analyst throughput. Cost avoidance is measured through prevented breach impact, reduced downtime, lower forensic and remediation costs, and reduced regulatory exposure. AI value becomes most persuasive when presented as a set of measurable operational improvements that directly connect to the business consequences executives already understand.

Competitive advantage and trust belong in the value narrative, but they must be grounded. Strong security is a market differentiator only when it is operationally real, demonstrably reliable, and communicated in a trustworthy way. AI contributes to competitive advantage when it improves service reliability, reduces customer-impacting incidents, and enables safer innovation. The language for executive leadership should emphasize reliability, resilience, and controlled acceleration of innovation. AI becomes meaningful to the board when it is positioned as an enterprise capability that protects growth rather than as a technology that merely expands the security toolkit.

The final step in bridging the gap is measurement discipline. Every AI initiative should be connected to two or three executive-level outcomes and tracked through a small set of business-facing indicators. This prevents the program from becoming a technical showcase and forces strategic alignment. It also gives executives what they need most: clarity on why the initiative exists, what it changes, how success is defined, and how risk is controlled.

Crafting a Compelling AI Vision and Roadmap

An AI strategy that is not expressed as a coherent vision will be interpreted as fragmented experimentation. Executive leadership will support AI in security when the program reads as a disciplined investment that increases resilience and reduces enterprise exposure. The AI vision should therefore function as a "north star" that defines the future security operating model and explains how AI supports the organization's long-term objectives. A credible vision does not promise perfection. It promises measurable progress toward a more adaptive, more resilient security posture.

A strong vision defines a target end-state in operational terms. The best visions describe how detection, response, and governance will function differently when AI is integrated.

238

The vision should describe improved speed, improved precision, and improved decision quality under uncertainty. The vision should also describe how human expertise is preserved and elevated rather than displaced. Executives are more likely to trust a vision that explicitly frames AI as augmentation, with humans retaining accountability and oversight.

The roadmap turns aspiration into a controlled plan. It should be staged, measurable, and risk-aware. Early phases should focus on foundations: data readiness, security telemetry quality, access controls, AI governance roles, vendor evaluation discipline, and pilot use cases that produce defensible learning. Middle phases should focus on operational integration: workflow embedding, playbook alignment, training and enablement, and performance tuning. Later phases should focus on scaling and resilience: cross-team adoption, expanded use cases, continuous monitoring, and formal validation through structured exercises and audits.

The roadmap must contain explicit decision points for leadership. Executives do not need model details, but they do need clarity on where the organization will pause, assess, and either continue, pivot, or stop. A mature roadmap includes checkpoints tied to outcomes and risk posture, not just timelines. These checkpoints should align with budget cycles and governance rhythms so AI does not become a perpetual pilot.

Resource transparency is part of credibility. The roadmap should clearly state what is required across technology, people, and process. This includes security engineering capacity, platform integration work, data management ownership, training investment, and operational readiness. A roadmap that ignores these realities will be perceived as optimistic and will lose leadership confidence when inevitable friction appears.

Finally, the roadmap must express success in business-aligned terms. Executives care about reduced response time, reduced incident impact, reduced operational disruption, and improved audit readiness. A roadmap that defines success using only technical metrics will be difficult to defend. A roadmap that defines success through business-level indicators becomes a governance instrument leadership can trust.

Presenting AI Risks and Mitigation Strategies Effectively

Executive buy-in depends on confidence, and confidence depends on credible risk communication. Presenting AI risk is not an exercise in caution for its own sake. It is an exercise in proving that the security function is adopting AI responsibly, with disciplined controls and clear accountability. Executives will support AI when they believe the organization understands the risk surface, can explain it plainly, and has the ability to manage it over time.

Algorithmic bias must be discussed in security terms, not academic terms. The executive concern is not fairness as an abstract principle. The executive concern is performance and exposure. Bias can create blind spots, increase false positives, and undermine trust in security decisions that affect employees, customers, or regulated outcomes. The mitigation approach must be concrete: dataset review and representativeness checks, controlled feature selection, model performance monitoring across meaningful segments, and a governance path for escalations when outcomes appear inconsistent or harmful.

Data privacy is a board-level risk because it is both regulatory and reputational. AI in security often touches high-volume telemetry that can include sensitive attributes. The mitigation strategy should be framed as layered controls: minimization, access control, encryption, retention discipline, and privacy-preserving techniques where applicable. The executive message should emphasize that privacy is not an afterthought and that AI does not expand data access without controls that are stronger than prior norms.

Adversarial manipulation is an AI-specific risk executives increasingly recognize. The language should be direct and non-alarmist: attackers can attempt to mislead models, poison training signals, or craft inputs designed to evade detection. Mitigation includes adversarial testing, input validation, ensemble approaches for critical detections, and monitoring for model drift or unexplained behavior changes. The key is to show executives that AI defenses are treated as attackable systems and engineered accordingly.

Over-reliance risk should be framed as operational fragility. If the workforce defers judgment to AI outputs, the organization loses resilience when models fail, drift, or encounter novel conditions. Mitigation requires training, clear human accountability, and workflow design that expects validation and feedback loops. The best executive framing positions human expertise as the control layer that makes AI safe at scale.

Risk communication is strengthened when it is paired with governance clarity. Executives need to know who owns AI risk decisions, what triggers escalation, how exceptions are handled, and how performance is validated. The security leader should present risks as managed exposures with defined controls, monitoring, and review cycles. This approach builds trust because it demonstrates maturity rather than optimism.

Demonstrating ROI and Measuring Success with Executive Level Metrics

AI value becomes durable when it is measurable in executive language. The goal is not to flood leadership with dashboards. The goal is to define a small set of indicators that show whether AI is improving resilience, reducing risk, and increasing operational effectiveness. AI projects lose sponsorship when success is described through technical metrics that do not connect to enterprise outcomes.

Operational efficiency can be measured through reduced analyst time spent on low-value triage, reduced false positives, improved alert fidelity, and improved throughput per analyst. Incident performance can be measured through Mean Time to Detect, Mean Time to Contain, and Mean Time to Respond. These metrics are meaningful only when they are contextualized. Leaders need to understand whether improvement is significant enough to reduce impact exposure, reduce downtime, or reduce the probability of an event becoming material.

Cost value should be framed through cost avoidance and impact reduction rather than simplistic headcount reduction narratives. Prevented breach impact is difficult to quantify perfectly, but leadership accepts defensible estimates when assumptions are transparent. The most credible approach is to model avoided impact based on historical incident costs, downtime exposure, and regulatory penalty ranges, and then tie AI contributions to measurable improvements such as reduced dwell time and reduced scope of compromise.

Security posture improvement should be treated as a trend rather than a single score. Executives respond well to visible movement: fewer critical incidents, reduced repeat incident patterns, improved control coverage, and reduced exposure windows. Compliance and audit readiness also matter at the board level, and AI value can be shown through reduced audit preparation burden, improved evidence quality, and faster policy adherence detection.

Finally, ROI measurement should reinforce governance discipline. It should be clear how success is reviewed, how initiatives are stopped when they do not deliver, and how new investments are justified. Executives will fund AI when measurement proves that the program is accountable and outcomes-driven rather than exploratory and indefinite.

Building Trust and Securing Buy-in for AI Initiatives

Trust is built through transparency, consistency, and demonstrated control. Executive leadership supports AI in security when they believe the program is governed, measurable, and aligned to enterprise objectives. Trust is eroded when AI is presented as inevitable, magical, or self-justifying. Buy-in is not earned through enthusiasm. Buy-in is earned through disciplined communication and visible progress.

Transparency requires balanced reporting. If false positives spike during early rollout, executives should hear it early, along with the root cause, mitigation plan, and expected timeline to stabilize. This does not weaken confidence. It strengthens confidence by demonstrating operational integrity. Transparency also includes clarity about what AI can and cannot do, especially under novel conditions, and what role humans play when uncertainty is high.

Consistency is achieved through cadence and predictability. Executives lose confidence when AI appears only during funding asks or crisis moments. A regular rhythm of briefings, aligned to governance cycles, builds familiarity and reduces perceived volatility. These updates should be short, structured, and anchored to outcomes and risks.

Buy-in is sustained when leaders feel the program is controllable. This means clear ownership, clear escalation triggers, and clear decision points. Executives should understand what is being deployed, how it is validated, what guardrails exist, and what would trigger a pause or rollback. This framing positions AI as a controlled capability, not a gamble.

Trust grows fastest when AI outcomes are tied to real operational stories supported by metrics. A single narrative of an AI-assisted detection that prevented widespread disruption can be powerful when paired with measured impact: time saved, scope limited, downtime avoided, and response accelerated. Storytelling is not a replacement for data. Storytelling is the method that makes data memorable and actionable.

Operationalizing Executive Communication

Most AI strategy communication fails not because the content is wrong, but because it lacks operational structure. Executive communication must be treated as a program capability with defined artifacts, a consistent cadence, and decision packages designed for leadership action. When this structure exists, AI strategy becomes governable rather than abstract.

The first requirement is a standard set of artifacts. A board-ready AI strategy packet should include a one-page executive summary, a simple roadmap view with phases and checkpoints, an outcomes-and-metrics panel, and a risk-and-controls panel. Each artifact should emphasize clarity over completeness. Technical details belong in appendices or in working sessions with technical leaders, not in board-level discussions.

The second requirement is cadence. A practical model is quarterly deep dives aligned to risk governance and budget cycles, supported by short monthly operating updates for senior leadership. The AI strategy should appear as part of enterprise risk and resilience, not as a separate technology track. This keeps it aligned to leadership priorities and prevents AI from being perceived as discretionary or experimental.

The third requirement is decision packaging. Executives make choices when options are presented clearly. Every major request should be expressed as a decision: proceed, pause, or pivot. Each option should include expected value, risk implications, resource requirements, and what the organization will measure in the next review cycle. This turns AI funding and expansion into accountable governance decisions rather than generalized optimism.

The fourth requirement is stakeholder alignment. Communications must be consistent across the Chief Information Officer, Chief Risk Officer, General Counsel, and business unit leadership. Misalignment between these voices will undermine trust regardless of technical success. The security leader should ensure that AI strategy messaging is coordinated, especially on privacy, accountability, and regulatory posture.

When executive communication is operationalized, AI strategy becomes easier to sustain. It becomes a managed program with shared language, measurable outcomes, and controlled risk. That operational maturity is often what separates AI initiatives that scale from AI initiatives that stall.

Board-Ready Brief

Communicating AI Strategy to Executive Leadership

Executive Summary

Artificial intelligence is rapidly reshaping the cybersecurity threat landscape and the defensive capabilities organizations deploy in response. However, AI delivers enterprise value only when it is clearly understood, governed, and measured at the executive level. Chapter 12 establishes a disciplined approach for communicating AI strategy in cybersecurity to senior leadership and boards, translating technical capability into business value, enterprise risk reduction, and measurable performance outcomes.

The chapter emphasizes that AI should not be presented as a standalone technology initiative, but as a strategic enabler of resilience, operational efficiency, regulatory confidence, and competitive advantage. Effective communication is the difference between AI becoming a sustained capability supported by leadership, or a fragmented technical experiment lacking governance and long-term impact.

Why This Matters to the Board

Cyber risk now represents a material business risk with financial, operational, regulatory, and reputational implications. AI is simultaneously increasing adversary sophistication and raising expectations for defense speed, accuracy, and scale. Boards are increasingly expected to oversee not only cybersecurity outcomes, but also the responsible use of AI in security operations.

Without a clear executive narrative, AI investments risk being misunderstood, underfunded, or misaligned with enterprise priorities. Chapter 12 addresses this gap by providing a framework that allows directors and senior executives to understand:

- What AI is changing in the security program
- How AI investments reduce enterprise risk and operational disruption
- How AI-related risks are governed and controlled
- How success is measured and reported over time

This approach enables informed oversight and confident decision-making rather than reactive approval of opaque technical initiatives.

Translating AI from Technical Capability to Business Value

A central theme of the chapter is the necessity of reframing AI discussions away from algorithms and models and toward outcomes leadership already governs. Executives do not need to understand how a model functions internally; they need clarity on what it changes for the business.

AI in cybersecurity creates value through five primary pathways:

1. **Risk Reduction**
   AI improves the organization's ability to detect threats earlier, prioritize what matters most, and limit attacker dwell time. Earlier detection and faster

containment directly reduce the likelihood of material incidents and the severity of those that occur.

2. **Operational Resilience**
   AI enables faster decision-making and coordinated response, reducing downtime, preserving service availability, and protecting critical operations during incidents.

3. **Efficiency and Cost Control**
   Automation and intelligent prioritization reduce analyst overload, lower false positive rates, and improve throughput without proportional increases in headcount.

4. **Regulatory and Audit Confidence**
   AI supports continuous monitoring, improved evidence collection, and more consistent enforcement of controls, strengthening compliance posture.

5. **Trust and Competitive Advantage**
   Strong, demonstrable security capabilities support customer confidence, protect brand reputation, and enable secure innovation.

Chapter 12 stresses that every AI initiative should be explicitly tied to one or more of these outcomes in executive communications.

Crafting a Clear AI Vision and Roadmap

Boards and executives require more than isolated success stories; they need a coherent strategy. The chapter outlines how to articulate an AI vision that is aspirational but grounded in business reality.

An effective AI vision for cybersecurity:

- Defines the desired future security posture (predictive, adaptive, resilient)
- Emphasizes augmentation of human judgment rather than replacement
- Aligns directly with enterprise risk appetite and growth objectives

This vision is operationalized through a phased roadmap that allows leadership to govern progress. The roadmap should clearly identify:

- Priority use cases aligned to business risk
- Phased implementation with decision checkpoints
- Resource requirements and skill dependencies
- Expected outcomes and executive-level metrics

Importantly, the roadmap is positioned as a living governance artifact, reviewed and adjusted as threats, technology, and business priorities evolve.

Communicating AI Risks with Credibility

Chapter 12 emphasizes that executive trust is built not by minimizing risk, but by addressing it transparently and proactively. AI introduces distinct risks that must be clearly articulated in business terms.

Key AI-related risk categories include:

- **Algorithmic bias and uneven performance**, which can create blind spots or unfair outcomes
- **Data privacy and regulatory exposure**, given the scale and sensitivity of data used by AI systems
- **Adversarial manipulation**, including attempts to deceive or poison AI models
- **Over-reliance on automation**, leading to skill atrophy and reduced human vigilance

The chapter provides guidance on presenting these risks alongside concrete mitigation strategies, such as data governance, validation controls, human oversight, adversarial testing, and clear accountability structures. This balanced presentation reinforces confidence rather than resistance.

Measuring ROI and Success with Executive-Level Metrics

For AI initiatives to sustain leadership support, their impact must be measurable. Chapter 12 establishes that success should be tracked using a small, stable set of metrics that align with executive oversight responsibilities.

Recommended executive-level indicators include:

- Mean Time to Detect, Respond, and Contain priority incidents
- Reduction in false positives and analyst workload strain
- Incident impact metrics, such as scope limited and downtime avoided
- Cost avoidance estimates tied to reduced dwell time and faster containment

- Compliance readiness indicators and audit efficiency improvements

The chapter cautions against overloading leadership with technical metrics and instead emphasizes trend-based reporting that demonstrates sustained improvement over time.

Building Trust and Securing Sustained Buy-In

Trust is the enabling condition for AI adoption at scale. Chapter 12 outlines how CISOs and security leaders can build and maintain executive trust through consistent, transparent engagement.

Key trust-building practices include:

- Regular, predictable communication cadence aligned to governance cycles
- Honest reporting of both progress and challenges
- Clear ownership and accountability for AI-related risk
- Visible learning and improvement over time

Executives are more likely to champion AI initiatives when they understand not only what is working, but how issues are identified, addressed, and governed.

Board Oversight Considerations

Directors should expect leadership to be able to answer the following consistently:

- How does AI materially reduce enterprise cyber risk?

- How is AI performance measured and improving over time?
- What AI-specific risks exist, and how are they controlled?
- Where is human accountability enforced in automated decisions?
- What decisions require board input at each roadmap phase?

These questions anchor AI oversight in governance rather than technology.

Executive Takeaway

AI in cybersecurity is no longer experimental; it is becoming foundational. However, its value is realized only when it is communicated, governed, and measured as a business capability. Chapter 12 provides a disciplined framework that enables leadership to understand AI's role, oversee its risks, and evaluate its return with confidence. When communicated effectively, AI becomes not just a defensive tool, but a strategic asset that strengthens resilience, protects trust, and supports long-term enterprise success.

Conclusion: Communicating AI Strategy

The success of artificial intelligence in cybersecurity is determined as much by how it is communicated and governed as by how it is engineered. AI introduces powerful capabilities, but without a clear executive narrative, disciplined oversight, and measurable outcomes, those capabilities risk becoming fragmented technical initiatives rather than sustained strategic assets. Chapter 12 demonstrates that effective communication is not an ancillary activity, but a core leadership responsibility that enables AI to deliver durable enterprise value.

Translating AI from technical complexity into business relevance requires reframing discussions around risk reduction, resilience, efficiency, and trust. Executive leadership and boards must be able to clearly understand how AI investments change the organization's exposure to cyber risk, improve operational continuity, and support regulatory and customer confidence. When AI is positioned in terms of outcomes leadership already governs, it becomes actionable rather than abstract, and sponsorship becomes informed rather than aspirational.

Equally important is the transparent presentation of AI-related risks and their mitigation. Algorithmic bias, data privacy exposure, adversarial manipulation, and over-reliance on automation represent real enterprise risks that demand explicit oversight. Addressing these risks openly, with clear controls and accountability, strengthens leadership confidence and reinforces responsible adoption. Trust is built not by minimizing uncertainty, but by demonstrating disciplined governance and continuous learning.

Measurement completes the communication cycle. Executive-level metrics that reflect detection speed, containment effectiveness, cost avoidance, and operational efficiency provide the evidence necessary to sustain investment and guide strategic decisions. These measures allow leadership to evaluate AI performance over time, recalibrate priorities, and ensure alignment with enterprise objectives. When AI success is visible and defensible, it becomes a durable component of the organization's risk management and growth strategy.

Ultimately, communicating AI strategy effectively positions cybersecurity leadership as a strategic partner to the business. It enables boards and executives to exercise informed oversight, make confident investment decisions, and integrate AI into the organization's broader resilience and trust narrative. In an era where AI increasingly shapes both threat and defense, clarity, transparency, and disciplined communication are what transform AI from a technical capability into an enduring enterprise advantage.

Chapter 13: The Future of AI in Cybersecurity Leadership

The future of cybersecurity leadership will be defined by how well leaders anticipate the next wave of AI capability and the second order risks that arrive with it. Artificial intelligence is changing the speed of both defense and offense, and this acceleration is pushing security organizations beyond traditional planning horizons. The challenge is no longer limited to selecting tools or expanding coverage. The challenge is guiding an enterprise through an era where decision cycles compress, automation expands, and adversaries exploit the same innovations defenders adopt. In this environment, leadership requires foresight, governance discipline, and a clear model for human authority in AI mediated operations.

This chapter examines the trends most likely to reshape cybersecurity in the near to mid-term and the leadership implications attached to each. Generative AI is expanding from content creation into defense simulation, response guidance, and synthetic training data. Autonomous security operations are moving from narrow automation into orchestrated action that can outpace human response windows. Federated learning enables cross organization model improvement while preserving data locality, creating new collaboration paths and new integrity risks. Post quantum preparation is forcing a re-evaluation of cryptographic foundations and the long-term durability of AI enabled trust mechanisms. The convergence of AI with IoT and high-speed edge networks is expanding the attack surface while demanding security decisions closer to where data is generated. The thread connecting these developments is the same. The future belongs to leaders who treat AI as a governed capability within an accountable operating model, not as a tool category.

Emerging AI Trends Impacting Cybersecurity

Generative AI for Defensive Advantage

Generative AI is expanding into core defensive functions, not as a novelty but as a force multiplier for scale and speed. One of the most immediate advantages is AI enabled security simulation. Generative models can produce realistic adversary behaviors and evolving intrusion sequences, allowing security teams to pressure test controls using scenarios that change over time rather than repeating static test cases. This strengthens detection engineering and exposes weak assumptions in playbooks, escalation paths, and tooling integration. It also improves operational readiness because teams rehearse against more realistic conditions.

Generative AI also accelerates the creation and refinement of security artifacts. Playbooks, triage guides, and response communications can be produced faster and tailored to situational context. The value comes from speed, but the risk comes from misplaced trust. A generated procedure can be plausible and still wrong for the environment. Leadership must ensure quality control mechanisms remain explicit, including required validation points, constrained scope for automated drafting, and post incident review that uses outcomes to improve prompt patterns and operational templates.

Synthetic data generation is another trend that will expand. Security teams often lack diverse training datasets due to privacy constraints and the scarcity of high fidelity examples of rare attack sequences. Generative AI can create realistic synthetic data that enables model training and testing without exposing sensitive records. This capability increases defensive learning speed, but it also demands discipline. Synthetic data must be clearly labeled, lineage tracked and separated from production telemetry to avoid contamination and false assumptions about threat prevalence.

Autonomous Security Operations

Autonomous security is evolving from automation of tasks to orchestration of action. The operational benefit is a reduction in time to detect and time to respond, especially for fast moving threats where human decision cycles cannot keep pace. Autonomous systems can isolate endpoints, adjust access conditions, update network controls, and initiate evidence collection within narrow time windows. This becomes critical in environments with distributed infrastructure, high alert volume, or constrained staffing.

Autonomy introduces leadership obligations that traditional automation did not. The question is not whether an autonomous action can be executed, but whether it should be executed without oversight. A self-directed containment action can prevent a breach, but it can also disrupt revenue systems, interrupt customer workflows, or trigger unsafe fail-states in industrial environments. Governance must define autonomy boundaries, escalation triggers, and fail-safe design. The most resilient approach adopts progressive autonomy, where low risk actions become autonomous first, validation metrics are monitored continuously, and higher consequence actions remain supervised until proven safe in the organization's operational reality.

Federated Learning and Privacy Preserving Collaboration

Federated learning shifts model improvement away from centralized data collection and toward distributed training where model updates move instead of raw data. This trend is significant for cybersecurity because it enables collaboration across endpoints, business units, and even organizations without forcing sensitive telemetry into shared repositories. The opportunity is stronger detection across broader behavioral diversity, improved coverage against emerging threats, and reduced privacy exposure.

This model also creates new integrity risks. A federated learning system is only as trustworthy as the updates it aggregates. Adversaries may attempt to poison the learning process by injecting corrupted updates, or by shaping local data to skew the global model. Security leaders must treat federated learning as a governed system, not a data science feature. Controls include provenance tracking for participating nodes, integrity checks for updates, anomaly detection on aggregated gradients, and clearly defined rules about which data sources are eligible to influence the model. The strategic advantage remains real, but it is only durable when trust mechanisms are built into the collaboration model.

Post Quantum Preparation and Durable Trust Foundations

Quantum computing is not yet a universal operational threat, but it is already shaping security roadmaps because cryptography is the foundation of trust. If cryptographic assumptions break, identity, confidentiality, integrity, and non-repudiation become unstable across nearly every control plane. Cybersecurity leaders must treat post quantum preparation as a long horizon resilience initiative rather than a last-minute migration.

The leadership work begins with inventory and dependency mapping. Leaders must understand where classical cryptography is embedded in systems, contracts, identity platforms, and vendor integrations. The next step is adopting post quantum transition planning, including hybrid approaches during the shift. AI systems also rely on trusted data flows, secure update channels, and authenticated inference pipelines. As cryptographic standards evolve, AI enabled security must evolve with them. The goal is not immediate replacement. The goal is to avoid irreversible technology debt and preserving the integrity of trust mechanisms over the next decade.

AI Convergence with IoT, Edge Computing, and High-Speed Networks

IoT expansion and edge processing are turning security into a distributed systems challenge at unprecedented scale. AI is essential for visibility and anomaly detection across millions of devices, many of which lack robust built in controls. High speed networks and low latency processing push decision making closer to the edge, which reduces response time but increases complexity in governance and monitoring. The enterprise attack surface becomes broader, more heterogeneous, and harder to secure through centralized policy enforcement alone.

Security leaders must adopt a systems view where AI governance scales across device classes, environments, and data sensitivity levels. Edge AI can enable fast detection and local containment, but it also creates dispersed models that must be monitored for drift, tampering, and inconsistent enforcement. Leaders should expect adversaries to exploit weak device ecosystems, insecure update mechanisms, and poorly governed model deployments. The defensive posture must combine AI enabled monitoring with strict lifecycle controls for devices, identities, firmware, and the models operating in edge contexts.

The Evolving Threat Landscape AI-Powered Attacks and Defenses

Adversaries are using AI to increase precision, scale, and adaptability. Social engineering is becoming more convincing through personalization at scale. Malware is becoming more dynamic through evasive behavior and automated targeting. Reconnaissance is becoming more efficient as AI tools reduce the cost of identifying exploitable pathways. These trends undermine security strategies that rely heavily on static indicators and signature driven detection.

Defense must evolve toward behavior based analytics, rapid correlation, and continuous validation. AI enabled detection and response will be central, but so will the human governance layer that decides how much autonomy is acceptable and what level of explainability is required. The organizations most likely to succeed will use AI to reduce noise, accelerate investigation, and compress response timelines while preserving clear escalation paths, human override authority, and transparent accountability.

AI and the Future of the Cybersecurity Workforce

AI will reshape workforce expectations, not by eliminating the need for professionals, but by changing what professional value looks like. Routine triage and repetitive analysis will be increasingly automated. Human work will shift toward higher judgment functions, including threat reasoning, detection strategy, model oversight, response leadership, and governance execution. New roles will expand, including AI security governance leads, AI model risk specialists, and practitioners who can translate operational security goals into effective AI workflows.

The workforce risk is not only skill gap. The workforce risk is overreliance. When teams defer judgment to AI outputs, critical thinking atrophies and operational fragility increases. Leaders must invest in AI literacy and in disciplined teaming models where analysts can question outputs, demand evidence, and understand model limitations. The future workforce is not a replacement story. The future workforce is a stewardship story where human expertise governs machine scale.

Ethical AI and Societal Impact in Security

AI security systems can unintentionally expand surveillance, amplify bias, and create accountability ambiguity if governance is weak. High volume monitoring can erode privacy if data minimization and retention limits are unclear. Behavioral analytics can produce discriminatory outcomes if models learn historical bias and treat correlation as intent. Autonomous response can create harm if it triggers disruptive actions without proportionality and oversight.

Cybersecurity leadership must embed ethical AI principles into operating practice. Fairness and bias evaluation must be routine, not exceptional. Transparency expectations must be defined, including explainability requirements for high consequence decisions. Accountability must remain human owned, even when action is automated. Ethical governance is not separate from security governance. In an AI driven era, ethical failure becomes security failure because it undermines trust and invites regulatory, legal, and reputational exposure.

Conclusion: Leading with Foresight in the AI-Driven Security Era

The future of cybersecurity leadership will be defined less by mastery of individual technologies and more by the ability to govern complexity at speed. Artificial intelligence is no longer a discrete capability added to security operations; it is becoming a structural force that reshapes how threats emerge, how defenses operate, and how decisions are made. As AI accelerates both offense and defense, leadership must shift from reactive control toward anticipatory design. The organizations that endure will be those whose leaders recognize AI as a strategic amplifier that demands intentional oversight, disciplined boundaries, and continuous adaptation.

This chapter has illustrated that the trajectory of AI in cybersecurity is not linear or benign. Generative systems will redefine preparedness, autonomous operations will compress response timelines, federated learning will challenge traditional data governance models, and post-quantum pressures will test the durability of trust itself. Each of these advances introduces not only opportunity, but new categories of risk that cannot be mitigated through tooling alone. The central leadership task is therefore not choosing which AI capabilities to deploy, but determining how much autonomy is acceptable, where human authority must remain absolute, and how accountability is preserved when machines act at machine speed.

Cybersecurity leaders must also confront the reality that AI reshapes the human dimension of security. The workforce of the future will not be smaller, but it will be fundamentally different. Value will shift from manual execution to judgment, interpretation, governance, and ethical stewardship. Human expertise will increasingly be measured by the ability to question AI outputs, contextualize risk, and guide automated systems rather than operate them directly. In this environment, leadership responsibility includes cultivating AI literacy, preserving critical thinking, and preventing over-reliance on automated confidence.

Equally important are the ethical and societal implications of AI-driven security. Surveillance capability, bias propagation, and opaque decision-making pose risks that extend beyond organizational boundaries. When trust is eroded, security itself is weakened. Responsible leadership requires embedding fairness, transparency, and proportionality into AI-enabled security operations from the outset. Ethical governance is not a constraint on effectiveness; it is a prerequisite for sustainable resilience in an AI-mediated world.

Ultimately, the future of AI in cybersecurity leadership is not about prediction, but preparedness. It is about designing organizations that can absorb uncertainty, adapt faster than adversaries, and make sound decisions under pressure. Leaders who succeed will treat AI not as an infallible authority, but as a powerful collaborator governed by human intent and organizational values. In doing so, they will transform AI from a source of acceleration risk into a durable strategic advantage, ensuring that security remains not only intelligent, but trustworthy in the years ahead.

Definitions

AI Ethics: The branch of ethics that addresses the moral implications and societal impact of artificial intelligence, focusing on principles of fairness, accountability, transparency, and beneficence.

Explainable AI (XAI): A set of AI tools and techniques that allow human users to understand and trust the results and output created by machine learning algorithms.

Generative AI: A type of artificial intelligence capable of generating new content, such as text, images, code, or synthetic data, often based on patterns learned from existing data.

Human-AI Collaboration: The synergistic interplay between human intelligence and artificial intelligence, where each leverages its strengths to achieve a common objective more effectively than either could alone.

Machine Learning (ML): A subset of AI that enables systems to learn from data and improve their performance on a specific task without being explicitly programmed.

Natural Language Processing (NLP): A field of AI that focuses on enabling computers to understand, interpret, and generate human language.

Operational AI: The deployment of AI technologies within the day-to-day operations of an organization, such as in cybersecurity, for automating tasks, enhancing decision-making, and optimizing processes.

Security Orchestration, Automation, and Response (SOAR): A category of security management tools that leverage AI and automation to help security teams manage and respond to cyber threats more efficiently.

Synthetic Data: Artificially generated data that mimics the statistical properties of real-world data, often used to train AI models without compromising privacy or requiring access to sensitive information.

Threat Intelligence Platform (TIP): A system that aggregates, enriches, and analyzes threat data from various sources, often enhanced by AI to provide actionable insights.

www.ingramcontent.com/pod-product-compliance
Lightning Source LLC
Chambersburg PA
CBHW031923190326
41519CB00007B/389